1
Modern
Mathematics
for Schools

Modern Mathematics for Schools

Second Edition

Scottish Mathematics Group

Blackie

Chambers

Blackie & Son Limited
Bishopbriggs · Glasgow G64 2NZ
Furnival House · 14–18 High Holborn
London WC1V 6BX

W & R Chambers Limited
11 Thistle Street · Edinburgh EH2 1DG

Designed by James W. Murray

International Standard Book Numbers
Pupils' Book
Blackie 0 216 89400 X
Chambers 0 550 75911 5
Teachers' Book
Blackie 0 216 89401 8
Chambers 0 550 75921 2

Printed in Great Britain by
Thomson Litho Ltd, East Kilbride, Scotland
Set in 10pt Monotype Times Roman

Scottish Mathematics Group

Members associated with this book

W. T. Blackburn
Dundee College of Education
Brenda I. Briggs
Formerly of Mary Erskine School for Girls
W. Brodie
Trinity Academy
C. Clark
Formerly of Lenzie Academy
D. Donald
Formerly of Robert Gordon's College
R. A. Finlayson
Allan Glen's School
Elizabeth K. Henderson
Westbourne School for Girls
J. L. Hodge
Madras College
J. Hunter
University of Glasgow
T. K. McIntyre
High School of Stirling
R. McKendrick
Langside College
W. More
Formerly of High School of Dundee
Helen C. Murdoch
Hutchesons' Girls' Grammar School
A. G. Robertson
John Neilson High School
A. G. Sillitto
Formerly of Jordanhill College of Education
A. A. Sturrock
Grove Academy
Rev. J. Taylor
St. Aloysius' College
E. B. C. Thornton
Bishop Otter College
J. A. Walker
Dollar Academy
P. Whyte
Hutchesons' Boys' Grammar School
H. S. Wylie
Govan High School
Only with the arithmetic section

R. D. Walton
Dumfries Academy

Preface

Book 1 of the original series *Modern Mathematics for Schools* was first published in July 1965. This revised series has been produced in order to take advantage of the experience gained in the classroom with the original textbooks and to reflect the changing mathematical needs in recent years, particularly as a result of the general move towards some form of comprehensive education.

Throughout the whole series, the text and exercises have been cut or augmented wherever this was considered to be necessary, and nearly every chapter has been completely rewritten. In order to cater more adequately for the wider range of pupils now taking certificate-oriented courses, the pace has been slowed down in the earlier books in particular, and parallel sets of A and B exercises have been widely introduced. The A sets are easier than the B sets, and provide straightforward but comprehensive practice; the B sets have been designed for the more able pupils, and may be taken in addition to, or instead of, the A sets. Occasionally a basic exercise, which should be taken by all pupils, is followed by a harder one on the same work; in such a case the numbering is, for example, Exercise 2 followed by Exercise 2B. It is hoped that this arrangement, along with the 'Graph Workbook for Modern Mathematics', will allow considerable flexibility of use, so that while all the pupils in a class may be studying the same topic, each pupil may be working examples which are appropriate to his or her aptitude and ability.

Each chapter is backed up by a summary, and by A and B revision exercises; in addition, cumulative summaries and exercises have been introduced at the end of alternate books. A new feature is the series of Computer Topics from Book 4 onwards. These form an elementary introduction to computer studies, and are primarily intended to give pupils some appreciation of the applications and influence of computers in modern society.

Books 1 to 7 provide a suitable course for many modern Ordinary Level and Ordinary Grade syllabuses in mathematics, including the University of London GCE Syllabus C, the Associated Examining Board Syllabus C, the Cambridge Local Syndicate Syllabus C, and the Scottish Certificate of Education. Books 8 and 9 complete the work for the Scottish Higher Grade Syllabus, and provide a good preparation for all Advanced Level and Sixth Year Syllabuses, both new and traditional.

Related to this revised series of textbooks are the *Modern Mathematics Newsletters* (No. 1, April 1971), the *Teacher's Editions* of the textbooks, the *Graph Workbook for Modern Mathematics*, the *Three-Figure Tables for Modern Mathematics*, and the booklets of *Progress Papers for Modern Mathematics*. These new Progress Papers consist of short, quickly marked objective tests closely connected with the textbooks. There is one booklet for each textbook, containing A and B tests on each chapter, so that teachers can readily assess their pupils' attainments, and pupils can be encouraged in their progress through the course.

The separate headings of Algebra, Geometry, Arithmetic, and later Trigonometry and Calculus, have been retained in order to allow teachers to develop the course in the way they consider best. Throughout, however, ideas, material and method are integrated *within* each branch of mathematics and *across* the branches; the opportunity to do this is indeed one of the more obvious reasons for teaching this kind of mathematics in the schools—for it is *mathematics* as a whole that is presented.

Pupils are encouraged to find out facts and discover results for themselves, to observe and study the themes and patterns that pervade mathematics today. As a course based on this series of books progresses, a certain amount of equipment will be helpful, particularly in the development of geometry. The use of calculating machines, slide rules, and computers is advocated where appropriate, but these instruments are not an essential feature of the work.

While fundamental principles are emphasized, and reasonable attention is paid to the matter of structure, the width of the course should be sufficient to provide a useful experience of mathematics for those pupils who do not pursue the study of the subject beyond school level. An effort has been made throughout to arouse the interest of all pupils and at the same time to keep in mind the needs of the future mathematician.

The introduction of mathematics in the Primary School and recent changes in courses at Colleges and Universities have been taken into account. In addition, the aims, methods, and writing of these books have been influenced by national and international discussions about the purpose and content of courses in mathematics, held under the auspices of the Organization for Economic Co-operation and Development and other organizations.

The authors wish to express their gratitude to the many teachers who have offered suggestions and criticisms concerning the original series of textbooks; they are confident that as a result of these contacts the new series will be more useful than it would otherwise have been.

Contents

Algebra

Geometry

Arithmetic

Notation

Sets of numbers

Different countries and different authors
give different notations and definitions
for the various sets of numbers.
In this series the following are used:

E The universal set

ϕ The empty set

N The set of natural numbers $\{1, 2, 3, \ldots\}$

W The set of whole numbers $\{0, 1, 2, 3, \ldots\}$

Z The set of integers $\{\ldots, -2, -1, 0, 1, 2, \ldots\}$

Q The set of rational numbers

R The set of real numbers

The set of prime numbers $\{2, 3, 5, 7, 11, \ldots\}$

Algebra

Algebra

An Introduction to Sets

$$\{1, 2, 3, 4\} \qquad \{ \square , \square , \diagdown , \diamond , \diagdown \} \qquad \{a,e,i,o,u\}$$

1 Description of a set

Mathematics started long ago when men learned to count. Over the centuries, ideas of number, measure and shape led to the development of the arithmetic, algebra and geometry which we study at school.

A basic idea is that of a *set*, which is simply a collection of objects. Although such a collection may contain several identical objects, we are interested only in the distinct objects in the collection.

The objects in a set are called the *members*, or *elements*, of the set.

Examples.—The members of your class; the present members of the House of Commons; the set of even numbers; the collection of numbers 2, 4, 13, 15; the set of capital cities of Europe.

For every set we must be able to say whether a given object is a member of the set or not; that is, the set must be clearly defined. When it is possible to list the members of a set we write down the list in curly brackets, separating the members by commas.

For example,

$\{a, e, i, o, u\}$ is the set of vowels in the English alphabet;

$\{1, 3, 5, 7, 9\}$ is the set of the first five odd numbers.

A set is often denoted by a capital letter, such as V.

For example, $V = \{a, e, i, o, u\}$, read 'V is the set a, e, i, o, u.'

Illustrations

Description	List
The set of the first four even numbers, or $A = \{$first four even numbers$\}$	$A = \{0, 2, 4, 6\}$

Description	List
The set of months whose names begin with M, or B = {months whose names begin with M}	B = {March, May}
W, the set of whole numbers	W = {0, 1, 2, 3, 4, ...}
P, the set of prime numbers	P = {2, 3, 5, 7, 11, ...}

The dots indicate that the list goes on indefinitely; we can read this ' and so on'.

Exercise 1A

Using the { } brackets notation, *list* the following sets opposite the given capital letters:

A = The set of the first six letters of the English alphabet.

B = The set of whole numbers greater than 10 but less than 15.

C = The set of days in the week.

D = The set of colours in the traffic lights.

E = The set of even numbers between 3 and 9.

F = The set of odd numbers between 10 and 20.

G = The set of days in the week beginning with T.

H = The set of months with 30 days.

I = The set of subjects you have in school today.

J = The set containing the smallest number and the largest number from 35, 53, 39, 51, 33.

K = The set of whole numbers less than 8.

L = The set of numbers on a clockface.

Now describe or list four more examples of sets (which must be clearly defined).

Exercise 1B

List the following sets as in Exercise 1A.

A = {last five letters in the English alphabet}

B = {whole numbers greater than 97 but less than 105}

C = {values of British decimal coins}

D = {numbers obtained by doubling 5, 7, 10, 14}

E = {letters which appear once only in the word MATHE-MATICS}

F = {odd numbers between 20 and 30}

G = {numbers which divide into 36 exactly}

H = {leap years between 1970 and 1990}

I = {squares of the first six whole numbers}

J = {the smallest and largest three-digit numbers that can be made from 5, 2, 3, 4, 6 using a digit once only in each number}

K = {the first seven prime numbers} (see list on page 4)

L = {the seven basic colours of the rainbow}

Can you describe in words the set shown in Figure 1?

Many illustrations of sets can be seen in magazines, e.g. sets of stamps, of cars, of football players. Make a classroom collection of sets like this.

Exercise 2A

Describe these sets in words:

1 {A, B, C, D, E}   2 {Saturday, Sunday}

3 {red, amber, green}   4 {spring, summer, autumn, winter}

5 {January, June, July}   6 {Paris, London, Rome, Moscow}

7 {kilogramme, gramme}   8 {5p coin, 10p coin, 50p coin}

9 {0, 2, 4, 6, 8, 10}   10 {11, 12, 13, 14}

11 {+, −, ×, ÷}   12 {123, 132, 213, 231, 312, 321}

Exercise 2B

1 List the set of numbers between 859 and 925 which have 4 in the units place.

2 Describe in words:

    a {0, 1, 2, 3, 4, 5, ...}   b {1, 3, 5, 7, ...}   c {0, 3, 6, 9, 12, ...}

3 Arrange the following shapes in sets according to their number of sides. Then describe each set in words:

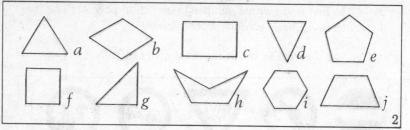

2

4 Form four sets from the following objects, and describe each set in words:

Monday, 4 pm, Thursday, 20p, 3 am, noon, £1·35, April, 50p, August.

5 $S = \{1, 2, 3, 4, 5\}$

  a List the set A whose members are formed by doubling each member of S.

  b List the set B of even numbers in S.

  c List the set C containing the smallest and largest numbers in S.

  d List the set D whose members are formed by multiplying each member of S by 5 and then subtracting 5.

6 $M = \{2, 4, 6, 8\}$. Form new sets as follows:

  a Set P, by adding 2 to each member of M.

  b Set Q, by multiplying each member of M by 12.

  c Set S, by halving each member of M.

  d Set T, by adding the elements of M two at a time in all possible ways.

7 List the set of numbers between 200 and 500 which have 5 in the tens place and 8 in the units place.

8 List the set of total scores possible when two dice are rolled.

2 Membership of a set $a \in \{a, b, c\}$

3

Each of the elements *a*, *b*, *c belongs to*, or *is a member of*, the set {*a*, *b*, *c*}. We show this by writing $a \in \{a, b, c\}$, $b \in \{a, b, c\}$ $c \in \{a, b, c\}$.

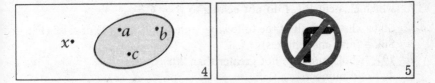

On the other hand, *x* does not belong to the set {*a*, *b*, *c*}, and we write $x \notin \{a, b, c\}$. Compare the stroke in the 'no right turn' road traffic sign.

Exercise 3

1   Use the symbol ∈ ('is a member of') to relate the following objects to their sets where possible. For example, cherry ∈ {fruit}.

Objects. Starling, giraffe, eagle, oak, spider, brick, pony, chestnut, apple, orange, sparrow, shark.

Sets. {birds}, {animals}, {fish}, {trees}, {fruit}.

2   State whether each of the following is true (T) or false (F):

a 5 is a member of the set of odd numbers.
b VI belongs to the set of Roman numerals.
c 215 ∈ {whole numbers divisible by 9}.
d Everest ∈ {mountains over 1000 metres high}
e Kilogramme ∈ {units of volume}.
f 100 ∈ {squares of the first ten whole numbers}.

3   Connect these elements with their sets, using '∈':

Elements: 2, cm, *a*, 4, *R*, Monday, *Q*, km, 6, *k*, Friday.
Sets: {1, 2, 3, 4}, {*a*, *b*, *c*, *d*}, {6}, {mm, cm, m, km}, {*Q*, *R*, *S*, *T*}, {days of the week}, {4, 5, 6}, {*k*, *l*}

4   $A = \{1, 2, 3, 4, 5\}$, $B = \{3, 6, 9, ..., 99\}$, $C = \{5, 10, 15, ...\}$.
Copy and complete the following by inserting the symbol ∈ or ∉ in place of the dots:

a 5 ... *A*   *b* 5 ... *B*   *c* 15 ... *B*
d 15 ... *C*   *e* 300 ... *B*   *f* 300 ... *C*

5   For the sets in question *4*,
 a   which element of *A* also belongs to *B*?
 b   which element of *A* also belongs to *C*?
 c   which elements of *A* do not belong to *B* or *C*?

6   State whether each of the following sentences is true (T) or false (F):
 a   Silk ∈ {man-made fibres}
 b   19 ∈ {whole numbers not greater than 20}
 c   4 ∉ {odd numbers}
 d   2008 ∉ {leap years}
 e   The Pole Star ∉ {planets of the Sun}
 f   2 ∉ {prime numbers}

---

3 The empty set = $\emptyset$ = { }

The empty set is the set with no members. It is denoted by the symbol ø or { }.

Examples of the empty set
1. The set of pupils in your class over 10 metres tall.
2. The set of whole numbers between 1 and 6 which are divisible by 8.

Can you think of some other examples of the empty set?
Do not confuse {0} with { }. {0} is a set which has one member, namely zero, and so is not empty.

Exercise 4

Which of the following are examples of the empty set? Write ø *or* { } if the answer is the empty set.

1   The set of men over 1000 years old.

2   The set of dogs which have ten tails.

3   The set of pupils in your class who are less than 6 years old.

4   The set of times when the hands of a clock are in a straight line.

5   The set of odd numbers which are divisible by 2.

6   The set of even numbers which are divisible by 5.

7   The set of prime numbers between 13 and 17.

8   The set of cubes with 10 sides.

9   The set of months that have more than 31 days.

10   The set of whole numbers between $\frac{1}{4}$ and $\frac{3}{4}$.

11   The set of straight lines that can be drawn on the curved surface of a cone.

12   The set of straight lines that can be drawn on the curved surface of a sphere.

13   The set of men who have reached the moon.

14   The set of months that start with the letter T.

15   The set containing the numeral 0 as its only member.

16   The set of pages in this book numbered 350 or over.

17   The set of even numbers which can be divided by 2 to give odd numbers.

18   The set of odd numbers which can be divided by 2 to give even numbers.

4 Equal sets have the same members

$$\{\triangle, \bigstar, \bigcirc\} = \{\bigcirc, \triangle, \bigstar\}$$

The *order* of listing the elements does not matter.

For example, if $A = \{a, b, c, d\}$ and $B = \{d, a, c, b\}$, then $A = B$.

Exercise 5

1   Find pairs of equal sets below, where possible, and write them like this: $\{p, q, r, s\} = \{s, p, r, q\}$.

$\{2, 4, 6\}$, $\{y, x\}$, $\{1, 3, 5, 7\}$, $\{x, y\}$, $\{1, 4, 9, 16\}$, {vowels in the English alphabet}, $\{2, 6\}$, $\{4, 2, 6\}$, $\{x, y, z\}$, {first four odd numbers}, $\{1 \times 1, 2 \times 2, 3 \times 3, 4 \times 4\}$, $\{u, e, i, o, a\}$.

2   Are the set $\{1, 5, 7, 9\}$ and the set of odd numbers less than 10 equal?

3   Which pairs of the following sets are equal?
     $A = \{0, 1, 2, 3, 4, 5\}$, $B = \{$whole numbers from 1 to 6 inclusive$\}$,
     $C = \{5, 3, 1, 0, 2, 4\}$, $D = \{$even numbers between 1 and 7$\}$,
     $E = \{$numbers on a die$\}$, $F = \{2 \times 1, 2 \times 2, 2 \times 3\}$.

4    $X = \{$letters in the word $\mathbf{REPEAL}\} = \{$R, E, P, A, L$\}$.
  a List the set Y of letters in the word $\mathbf{PARALLEL}$.
  b Is $X = Y$?

5   Write down three examples of pairs of equal sets of your own choice.

5 Sets within sets—subsets $B \subset A$

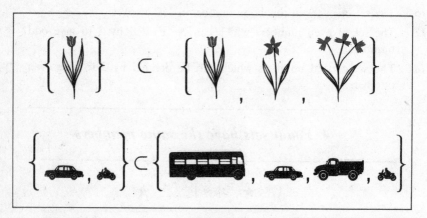

A set B is a *subset* of a set A if every member of B is a member of A.
We write $B \subset A$, which is read 'B is a subset of A'. We also say that
B is *contained in* A.

Example 1.—If $P = \{$pupils in your class$\}$ and
                 $M = \{$pupils in your class who have birthdays
                        in March$\}$,
M is a subset of P, i.e. $M \subset P$.

Can you suggest other subsets from the set of pupils in your class?
    Note.—The set of pupils in your class who take mathematics is P
itself, and the set of pupils in your class under 6 years of age is the

empty set ø. Hence, the set *P* itself and the empty set are also taken to be subsets of *P*,

$$\text{i.e. } P \subset P \text{ and } \varnothing \subset P.$$

Example 2.—By choosing as many elements as we please from the set $A = \{a, b\}$, we obtain all the possible subsets of *A* which are:

$$\{a\}, \{b\}, \{a, b\}, \varnothing.$$

Exercise 6A

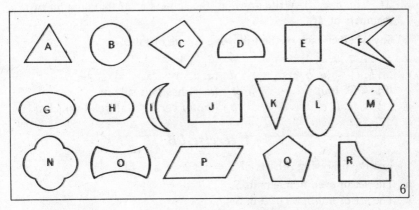

1   Figure 6 shows a set *S* of shapes with straight and curved sides. Use the capital letters to list the members of the following subsets of *S*:

a The subset of members of *S* with straight sides only.
b The subset of members of *S* with exactly three straight sides.
c The subset of the members of *S* with exactly four straight sides.
d The subset of the members of *S* with more than four straight sides.
e The subset of the members of *S* with curved sides only.
f The subset of the members of *S* with only one curved side.
g The subset of the members of *S* with exactly two curved sides.
h The subset of the members of *S* with more than two curved sides.
i The subset of the members of *S* with some straight and some curved sides.
j The subset of the members of *S* with eight straight sides.

2   List the following subsets of $A = \{1, 3, 5, 7, 9\}$:

a The set of prime numbers in *A*.
b The set of numbers in *A* divisible by 3.

c The set of numbers in A divisible by 2.
d The set of numbers in A greater than 5.
e The set containing the pair of numbers in A whose sum is 14.
f The set of numbers in A less than 1.

3   List all the subsets of the set $S = \{1, 2\}$

4   Write down some subsets of the set of positions players may have in a football team, or a hockey team, or another sports team.

5   $C = \{$John, Mary, Betty$\}$. Write down any four subsets of C.

6   $M = \{a, b, c, d\}$. Write down all the subsets of M that contain two members of M.

7   Say whether each of the following is true or false:

a $\{2, 3\} \subset \{2, 3, 4\}$ b $\{2, 3, 4\} \subset \{2, 3\}$
c $\{a, b, c\} \subset \{a, b, c\}$ d $\{a, b, c\} \subset \{x, y, z\}$
e $\{0\} \subset \{0, 100\}$ f $\{x, y\}$ has four subsets
g $\{ \ \} \subset \{a, b, c\}$ h $\{$motor cars$\} \subset \{$all road vehicles$\}$

Exercise 6B

1   List the following subsets of $S = \{1, 2, 3, 4, 5, 6\}$:
a The set of even numbers in S.
b The set of multiples of 3 in S.
c The set of multiples of 10 in S.
d The set of numbers in S which divide exactly into 48.
e The set consisting of three members of S whose sum is 15.

2   $A = \{$vowels$\}$, $B = \{$even numbers greater than 100$\}$, $C = \{a\}$, $D = \{1000\}$, $E = \{$vowels in the word *addition*$\}$, $F = \{1000, 1001\}$. Write down pairs of sets from the above, one of which is a subset of the other, e.g. $C \subset A$.

3   Describe in words some subsets from $\{1, 2, 3, 4, 5, 6, 7, 8\}$.

4   $S = \{3, 4, 5, 6, 7, 8\}$, $T = \{5, 7, 9, 11\}$. List the following subsets of S:
a Its members are greater than 8.
b Its members are divisible by 3.
c Its members are those members of S which are also members of T.
d Its members are members of S which are 3 less than members of T.

5   List all the subsets of each of the following. Remember that $\{p\} \subset \{p\}$ and $\emptyset \subset \{p\}$.

a $\{p\}$ b $\{p, q\}$ c $\{p, q, r\}$ d $\{p, q, r, s\}$

6 Use the results of question 5 to copy and complete this table:

Set	Number of members in set	Number of subsets
$\{p\}$		
$\{p, q\}$		
$\{p, q, r\}$		
$\{p, q, r, s\}$		
$\{p, q, r, s, t\}$		

7 Draw sketches of a cuboid, a cylinder, a cone and a sphere.
If $A = \{$cuboids, cylinders, cones, spheres$\}$, list the following subsets of A:

a The set of solids which have 6 faces.

b The set of solids which have no corners.

c The set of solids which have two curved surfaces.

d The set of solids which have at least one flat surface.

8 By forming suitable subsets A, B, C of your own choice from $\{1, 2, 3, 4, 5, 6, 7, 8, 9, 10\}$, find out whether or not the following sentence is true for the subsets chosen:
'If $A \subset B$ and $B \subset C$, then $A \subset C$.'

9 Which of the following are true?

a	$2 \in \{1, 2, 3\}$	*b*	$2 \subset \{1, 2, 3\}$	*c*	$\{2\} \in \{1, 2, 3\}$
d	$\{2\} \subset \{1, 2, 3\}$	*e*	$\{2\} \subset \{2\}$	*f*	$2 \in \{2\}$

---

6 The universal set E ('Entirety')

A *universal set* contains all the elements being discussed. It is denoted by the symbol E ('Entirety'). All other sets in the discussion will be subsets of E.

Example 1.—A universal set for $\{0, 2, 4, 6\}$ could be $\{$even numbers$\}$, or $\{$even numbers less than 8$\}$, or $\{$whole numbers$\}$.

Example 2.—Suitable universal sets for $\{$makes of cars beginning with V$\}$ could be $\{$makes of British cars$\}$, or $\{$makes of European cars$\}$.

Exercise 7

State a possible universal set for each of the following:

1   {a, b, c, d}   2   {cats, dogs, budgerigars}

3   {Spain, France, Holland}   4   {1, 3, 5, 7, 9}

5   {Mercury, Venus, Mars, Earth}   6   {copper, silver, gold}

7   {carrots, turnips, lettuce}   8   {pupils in your class}

9   {5p coin, 10p coin}   10 {red, yellow, black, blue}

Give *two* possible universal sets for each of the following:

11 {3, 5, 7}   12 {a, e, i}

13 {residents of Edinburgh}   14 {all British cars}

15 {3, 6, 9, 12}   16 {64, 128, 9652}

17 {cube, cuboid}   18 {51, 52, 53, 54, 55}

---

7 Pictures of sets—Venn diagrams

There is a useful method of representing sets by a picture. All the members of the universal set are represented by dots marked inside a rectangle, each dot being suitably labelled as in Figure 7, which represents $E = \{1, 2, 3, 4, 5, 6, 7, 8, 9\}$.

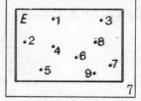
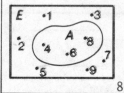
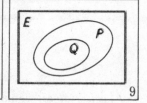

A subset of the universal set is shown by drawing a closed curve around the dots representing its members. Figure 8 shows the subset $A = \{4, 6, 8\}$ of E.

When the number of elements in each set is large the dots are omitted from the diagram. For example, Figure 9 shows the diagram for $E = \{$pupils in your town$\}$, $P = \{$pupils in your school$\}$, $Q = \{$pupils in your class$\}$.

The use of colours or shading gives clearer diagrams. Such diagrams are called Venn diagrams, after the English mathematician John Venn who lived from 1834 to 1923.

Exercise 8A

1   Draw a Venn diagram to show each of the following, marking in labelled dots to represent the members of the sets:

a $E = \{$vowels in the English alphabet$\}$; $A = \{$the first three vowels$\}$

b $E = \{$blue, green, red$\}$; $B = \{$blue$\}$

c $E = \{$whole numbers from 1 to 9$\}$;
$C = \{$even numbers between 1 and 9$\}$

2   Illustrate each of the following by a Venn diagram:

a $E = \{$all roses$\}$; $W = \{$white roses$\}$

b $E = \{$pupils in your class$\}$;
$G = \{$pupils in your class who wear glasses$\}$

c $E = \{$whole numbers$\}$; $A = \{$odd numbers$\}$

3   Draw a Venn diagram for the set E of pupils in your school and the subset A of pupils in your class.
Now add subset B of pupils in your class who are left-handed; think carefully where to show this.

4   Illustrate each of the following by a Venn diagram:

a $E = \{1, 2, 3, 4, 5, 6\}$; $P = \{1, 2, 3\}$, $Q = \{4, 5\}$

b $E = \{a, b, c, d, e, f\}$; $F = \{a, b, c, d\}$, $G = \{c, d, e\}$

c $E = \{0, 1, 2, 3, 4\}$; $M = \{1, 2, 3\}$, $N = \{2\}$

5   From Figure 10, list the members of the following sets:

a   Set A **b**   Set B

c   The set of elements that belong to both A and B.

d   The set of elements that belong to either A or B (or both A and B).

e   The set of elements that do not belong to A or B.

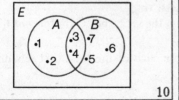

10

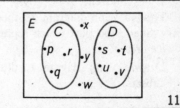

11

6   From Figure 11 on page 15, list the members of the following sets:
a   Set *C*   b   Set *D*
c   The set of elements that belong to both *C* and *D*.
d   The set of elements that belong to either *C* or *D* (or both *C* and *D*).
e   The set of elements that do not belong to *C* or *D*.

7   In Figure 12, *E* = {pupils selected from a senior class};
     H = {pupils who like history} and
     G = {pupils who like geography}. Each letter represents a pupil.

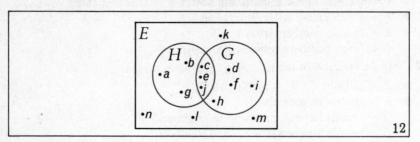

12

a   How many pupils like history?
b   How many pupils like geography?
c   How many pupils like both history and geography?
d   How many pupils like history but not geography?
e   How many pupils like geography but not history?
f   How many pupils like neither history nor geography?
g   How many pupils are in set *E* altogether?

Exercise 8B

1   *E* is the set of squares of the first seven whole numbers. *A* is the set
     of numbers between 11 and 39 which have 6 as units digit and which
     are also divisible by 4. List the sets *E* and *A*.
     Illustrate by a Venn diagram.

2   *E* = {children}, *B* = {boys}, *C* = {boys who like ice-cream}.
     Draw a Venn diagram for these sets.
        Shade the part which represents the set of boys who do *not* like
     ice-cream.

3   Draw a Venn diagram to illustrate each of the following sets and
     subsets:

  a $E = \{$all road vehicles$\}$; $A = \{$all buses$\}$, $B = \{$all lorries$\}$.

  b $E = \{$all musical instruments$\}$; $C = \{$all stringed instruments$\}$,
     $D = \{$all violins$\}$.

  c $E = \{$pupils in your school$\}$; $G = \{$pupils who wear glasses$\}$,
     $L = \{$pupils who are left-handed$\}$. [Assume that there are some left-handed pupils who wear glasses.]

  d $E = \{$natural numbers$\}$;
     $P = \{$natural numbers less than 40 which are divisible by 5$\}$,
     $Q = \{$natural numbers less than 40 which end in 5 or 0$\}$.
     What can you say about P and Q?

4 $E = \{$all British subjects$\}$;
     $K = \{$British subjects under 16 years of age$\}$,
     $M = \{$British subjects with a driving licence$\}$.

     Draw a Venn diagram to show that you must be 16 years or over to hold a driving licence.

5   Which of these Venn diagrams illustrate the pairs of sets given below, where E is the set of whole numbers?

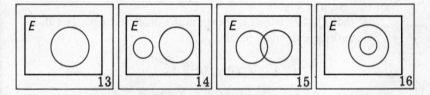

  a $A = \{$even numbers$\}$, $B = \{$odd numbers$\}$.

  b $A = \{$even numbers$\}$, $P = \{$prime numbers$\}$.

  c $D = \{$multiples of 2$\}$, $F = \{$multiples of 4$\}$.

  d $S = \{$numbers divisible by 2 and by 5$\}$,
     $T = \{$numbers divisible by 10$\}$.

  e $D = \{$multiples of 2$\}$, $K = \{$multiples of 3$\}$.

8 Intersection of sets $A \cap B$

The *intersection* of two sets A and B is the *set* of elements which are members of A *and also* members of B.

This new set is written $A \cap B$ ('A intersection B').

Illustration 1. If $A = \{p, q, r, s\}$ and $B = \{r, s, t\}$, then
$$A \cap B = \{r, s\}.$$
The shaded region in the Venn diagram in Figure 17 represents $A \cap B$.

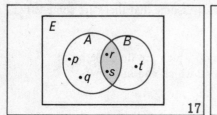 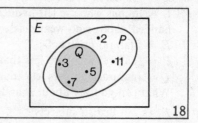

Illustration 2. If $E = \{\text{whole numbers}\}$,
$P = \{\text{prime numbers less than 12}\}$,
$Q = \{\text{odd numbers between 2 and 8}\}$,
list P, Q, $P \cap Q$ and illustrate by a Venn diagram.
$P = \{2, 3, 5, 7, 11\}$, $Q = \{3, 5, 7\}$, $P \cap Q = \{3, 5, 7\}$.
The intersection of P and Q is shown by the shaded region in the Venn diagram in Figure 18.

Note that $Q \subset P$, which is shown in the diagram by the fact that set Q is entirely contained in set P.

Exercise 9A

1   $A = \{\text{first five odd numbers}\}$, $B = \{7, 9, 11\}$, $C = \{1, 2, 3, 4\}$. List $A \cap B$, and $A \cap C$.

2   $D = \{\text{prime numbers less than 12}\}$, $F = \{\text{even numbers less than 12}\}$, $W = \{\text{whole numbers}\}$. List $D \cap F$, and $D \cap W$.

3   $E = \{\text{letters of the alphabet}\}$; $P = \{a, b, c, d, e, f\}$, $Q = \{b, c, d\}$, $R = \{d, e, f\}$, $S = \{f\}$.
    List the sets $P \cap Q$, $Q \cap R$, $R \cap S$, $P \cap R$, $Q \cap S$.

4   $E = \{1, 2, 3, 4, 5, 6\}$; $A = \{1, 2, 3, 4\}$, $B = \{3, 4, 5\}$, $C = \{5\}$.
    a List the sets $A \cap B$, $B \cap C$, $A \cap C$, $E \cap B$.
    b Show these intersections in separate Venn diagrams. The use of shading or colouring will help.

5   $V = \{\text{vowels}\}$, $H = \{h, o, l, i, d, a, y, s\}$, $E = \{\text{letters of the alphabet}\}$.
    Illustrate $V \cap H$ by a Venn diagram.

6   $X = \{\text{even numbers less than 9}\}$, $Y = \{\text{odd numbers less than 8}\}$, $Z = \{\text{numbers between 1 and 10 which are divisible by 3}\}$.

a List the sets *X*, *Y* and *Z*.

b List the sets $X \cap Y$, $Y \cap Z$, $X \cap Z$. Show these intersections in Venn diagrams with the set of whole numbers as universal set.

7 Copy these diagrams and shade the regions that represent the given intersections:

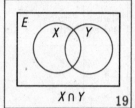

$X \cap Y$ 19

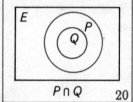

$P \cap Q$ 20

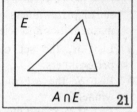

$A \cap E$ 21

8 In a certain class, these girls studied Art—Betty, Freda, Wilma, Carol, Lynn; and these girls studied Music—Pat, Carol, Lynn, Sue.

a Make a Venn diagram to illustrate this information.

b Shade the region which represents the set of girls who studied both Art and Music.

9 In a group of boys, 15 like ice-cream, 20 like crisps, and 12 like both.

a Illustrate these facts in a Venn diagram.

b How many boys are in the group?

10 Of 25 girls, 18 like pop music, 13 like classical music, and 10 like both.

a Illustrate these facts in a Venn diagram.

b How many girls like neither pop nor classical music?

Exercise 9B

1 A = {whole numbers between 1 and 19 which are divisible by 3},
B = {whole numbers between 1 and 19 which are divisible by 2}.

a List *A* and *B*.

b List the set of members of *A* which are also members of *B*.

c List the set of members of *B* which are also members of *A*.

d Is $A \cap B = B \cap A$?

2 P = {factors of 24} = {1, 2, 3, 4, 6, 8, 12, 24}.

a List the set *Q* of factors of 36.

b List $P \cap Q$.

c Hence state the largest factor of both 24 and 36.

3     $V = \{$vowels$\}$, $F = \{f, a, c, e, t, i, o, u, s\}$. Show $V \cap F$ in a Venn diagram.

4   In your class, $J = \{$pupils whose first names begin with $J\}$, and $S = \{$pupils whose surnames begin with $S\}$.
    List sets J, S and $J \cap S$, and show these in a Venn diagram.

5     $P = \{a, b, c, d, e\}$ represents the set of prizewinners in a school for mathematics and science:

Pupil	Maths. Prize	Science Prize
a	–	×
b	×	×
c	×	–
d	×	×
e	×	–

   a   List M, the set of mathematics prizewinners.

   b   List S, the set of science prize-winners.

   c   List $M \cap S$, and illustrate in a Venn diagram.

6   There are 20 boys in a certain class. 16 of them study physics, 14 study chemistry and 12 study both physics and chemistry.

   a   Draw a suitable Venn diagram.

   b   Deduce the number of boys in the class who study neither physics nor chemistry.

7   In a group of 16 girls, 10 like lemonade, 12 like potato crisps and 8 like both.
    Draw a Venn diagram, and find how many like neither lemonade nor crisps.

8   Figure 22 shows a typical Venn diagram for three sets A, B and C. The *intersection of the three sets*, $A \cap B \cap C$, is shown by the shaded region, and is the set of elements which are members of A, B and C. Simplify:

   a $\{1, 3, 5, 7\} \cap \{3, 5, 8\} \cap \{3, 9\}$

   b $\{p, q, r\} \cap \{p, r, s, t\} \cap \{m, n, p, r, s\}$

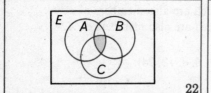

22

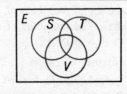

23

9   In Figure 23, E represents the set of all pupils in your town, S the
set of pupils in your school, T the set of pupils whose surnames start
with T, V the set of pupils who are 12 years old.
Describe each of the following sets *in words*:

    a $S \cap T$ *b* $S \cap V$ *c* $T \cap V$ *d* $S \cap T \cap V$.

10 $E = \{1, 2, 3, 4, 5, 6, 7, 8, 9, 10\}$. $A = \{2, 3, 4, 5, 6\}$,
$B = \{3, 5, 6, 7, 8, 9\}$ and $C = \{4, 5, 6, 9, 10\}$ are subsets of E.

  a List the sets $A \cap B$, $B \cap C$, $C \cap A$, $A \cap B \cap C$.
  b Illustrate by a Venn diagram.

Exercise 10 Miscellaneous Examples

1   State whether each of the following is true or false:

  a The set of pupils under 10 years old in your class is the empty set.
  b $5\frac{1}{2}$ is a member of the set of whole numbers.
  c If $A = \{v, x, z, k, q\}$ and $B = \{x, z, p\}$ then $B \subset A$.
  d For sets A and B in part *c*, $A \cap B = \{x, z\}$.
  e $11 \in \{$prime factors of $306\}$.
  f Every set has at least one member.
  g If A is the set of vowels in the alphabet and B is the set of all letters
in the alphabet, $A \subset B$.
  h If C and D are sets, $C \cap D = D \cap C$.
  i If $A \subset B$ and $a \in A$, then $a \in B$, where A and B are sets.
  j If $X \subset Y$ then $X \cap Y = X$, where X and Y are sets.

2   Using the symbols $=$, $\cap$, $\subset$, and $\emptyset$, make as many statements as
you can about the following pairs of sets;
for example, if

$$P = \{a, e, i, o, u\} \quad \text{and} \quad Q = \{u, a, o, e, i\}$$

    then $P = Q$, $P \cap Q = P = Q$, $P \subset Q$, $Q \subset P$

  a $A = \{1, 2, 3, 4, 5\}$, $B = \{3, 4, 5, 6\}$
  b $C = \{5, 10, 15, 20\}$, $D = \{10, 15\}$.
  c $F = \{a, b, c, d\}$, $G = \{e, f, g, h\}$.
  d $H = \{0, 1, 2\}$, $K = \{2, 0, 1\}$.

3   Draw Venn diagrams to illustrate the following, and in each case
state an appropriate universal set:

  a $A = \{a, b, c, d, e\}$, $B = \{b, c, g\}$.
  b $C = \{$whole numbers from 1 to 9 inclusive$\}$, $D = \{3, 6, 9\}$.

c F = {multiples of 3, less than 18}, G = {multiples of 5, less than 20}.

d H = {odd numbers}, K = {even numbers}.

4 $X \subset Y$. What can you say about the sets X and Y if $Y \subset X$ also? What does $X \cap Y$ equal? (Two answers.) Give an example to illustrate your answers.

5 A = {M, A, T, H, S}, and B = {M, A, T, H}. Which of the following statements are true?

$$A \subset B, \quad B \subset A, \quad A = B, \quad A \cap B = B, \quad B \in A, \quad A \in B.$$

6   If you know that $a \in X$ and $a \in X \cap Y$, can you make any other statement about a?

7   If $A \subset B$ and $A \cap C = \emptyset$ draw a Venn diagram showing that $C \subset B$ is possible.

8   If there are 10 members in the set A, 7 members in the set B, and 4 members in the set $A \cap B$, how many members of A are *not* members of B?

9   Here are four pairs of sets, each contained in the universal set of the whole numbers:

a   The set of multiples of 5; the set of multiples of 10.

b   The set of multiples of 3; the set of multiples of 2.

c   The set of numbers divisible by 3 and also by 5; the set of numbers divisible by 15.

d   The set of odd numbers; the set of multiples of 8.

Check that the four diagrams given below correctly illustrate the above pairs of sets. Pair off *a*, *b*, *c*, *d* with (i), (ii), (iii), (iv).

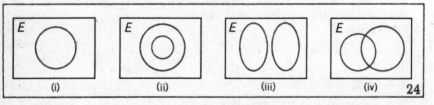

(i)   (ii)   (iii)   (iv)   24

10   Jack delivers morning newspapers; 35 of his customers take both a daily and a Sunday paper, 10 take only a daily paper and 15 take only a Sunday paper. How many customers does Jack have in all?

Summary

1 A *set* is a clearly defined collection of objects.

$A = \{a, b, c\}$

2 Each object in a set is a *member* or *element* of that set, and *belongs* to the set.

$a \in A, p \notin A$

3 *Equal sets* have exactly the same members.

$B = \{c, a, b\}, A = B$

4 The *empty set* is the set with no members.

$\emptyset$ *or* { }

5 A *universal set* is a set containing all the elements being discussed.

E ('Entirety')

6 A set B is a *subset* of a set A if every element of B is a member of A.

$B \subset A, A \subset A$
$\emptyset \subset A, \emptyset \subset E$.

7 The *intersection* of two sets A and B is the set of elements which are members of A and are also members of B.

$A \cap B$

8 *Venn diagrams*:

$A \cap B$

$A \subset B$

$A \cap B = \phi$

$A = B$

Mathematical Sentences: Equations

1 True sentences and false sentences

The information given in a sentence may make the sentence true or false.

For example, 'The sum of 5 and 18 is 23' is *true*,

but 'The product of 5 and 8 is 41' is *false*.

Exercise 1A

Say whether each of the following sentences is true (T) or false (F).

1 There are 1000 pence in £10.

2 7 is an even number.

3 A cube has 8 corners.

4 $8 \times 0 = 8$.

5 895 is divisible by 5.

6 Five months of every year have exactly 30 days.

7 There are only 3 prime numbers between 10 and 20.

8 A 50-pence British coin has 7 sides.

9 The metre is a unit of length.

10 6 is a factor of 40.

11 {1, 2} has 4 subsets.

12 The hour hand of a clock turns round 12 times between midnight one day and midnight the next day.

Exercise 1B

Say whether each of the following sentences is true (T) or false (F).

1 A cuboid has 8 faces.

2 8 is a factor of 72.

3 {1, 2, 3} has 8 subsets.

4   The year 2984 will be a leap year.

5   The answers to $3+1\cdot5$ and $3\times1\cdot5$ are not the same.

6   9260 is divisible by 5.

7   The Atlantic Ocean bounds the east coast of the United States of America.

8   There are exactly 3 prime numbers less than 10.

9   1 kg of feathers weighs less than 1 kg of lead.

10 There are 3600 seconds in 1 hour.

11 The minute hand of a clock turns round once in 24 hours.

12 581 is a prime number.

---

2 Open sentences

Look at the sentence '□ is a divisor of 12'. We cannot tell whether this sentence is true or false until □ is replaced by a natural number.

Thus if □ is replaced by 5, the sentence is false, but if □ is replaced by 4, the sentence is true.

A sentence like '□ is a divisor of 12', which is neither true nor false as it stands, is called an open sentence. A letter of the alphabet is often used in such a sentence instead of a shape; for example 'x is a divisor of 12'.

Illustrations

Open sentence	True sentence	False sentence
$x+2 = 7$	$5+2 = 7$	$4+2 = 7$
$y \in$ {even numbers}	$8 \in$ {even numbers}	$3 \in$ {even numbers}
z is less than 6	5 is less than 6	7 is less than 6

To show that □ is replaced by 4, we write □ $= 4$; 4 is a *replacement* for □.

Exercise 2

For questions *1-8* write down the headings *open sentence*, *true sentence*, and *false sentence* as shown in the illustrations above.

Copy down the given open sentence, and fill in *one* true sentence and *one* false sentence for each:

1    $x+5 = 8$ 2 $x-2 = 1$

3    $y \in$ {odd numbers}   4 z is greater than 10

5    $a \in$ {prime numbers}   6 b is a multiple of 7

7   c is divisible by 10   8 d is greater than one million

For questions *9–18*, give a replacement for each of the symbols $\square$, $\triangle$, $*$, x, a, p, z which changes the open sentence into a true sentence. Choose the replacements from the set of natural numbers {1, 2, 3, 4, ...}.

9    $\square + 3 = 5$; $\square = ...$ 10 $\square - 1 = 6$; $\square = ...$

11   $\triangle + 4 = 9$; $\triangle = ...$ 12 $2 \times \triangle = 8$; $\triangle = ...$

13   $*$ is less than 10; $* = ...$ 14 $*$ is a prime number; $* = ...$

15   $x+2 = 12$; $x = ...$ 16 $a-5 = 1$; $a = ...$

17   $z \div 4 = 9$; $z = ...$ 18 $4 \times p = 4$; $p = ...$

For questions *19–26*, write out the *set of replacements* for the symbols which changes each of the open sentences into a true sentence. Choose the replacements from the set of natural numbers.

For example, for the open sentence 'x is a divisor of 12', the set of replacements for x is {1, 2, 3, 4, 6, 12}.

19   y is a divisor of 15   20 $\square + 12 = 20$

21   $\triangle$ is less than 8   22 $2 \times p = 30$

23   $x-5 = 15$ 24 m is a multiple of 4, less than 21

25   $t \times t = 36$ 26 n is greater than 5 and less than 10

---

3 Solution sets

Each of the open sentences above contained a symbol (letter or shape) which was to be replaced by an element, or by elements, from the set of natural numbers. A symbol which may be replaced by any member of a given set is called a *variable* on that set.

Consider again the open sentence 'x is a divisor of 12', where x is a variable on the set of natural numbers.

When x is replaced by the elements 1, 2, 3, 4, 6, or 12, 'x is a divisor of 12' is made into a *true* sentence. These are the only such replacements.

Each of these replacements is said to *satisfy*, or to be a *solution* of the open sentence, e.g. $x = 3$ satisfies or is a solution of 'x is a divisor of 12'. The set of all solutions is called the *solution set* (or the *truth set*) of the open sentence.

{1, 2, 3, 4, 6, 12} is the solution set of 'x is a divisor of 12'.

If no such replacement can be found, the solution set is empty. Thus, for the same universal set, the solution set of the open sentence 'x is less than 1' is ø.

Exercise 3

In each of the open sentences *1-6*, x is a variable on

$$N = \{1, 2, 3, 4, ...\}.$$

Give the replacement for x which changes the open sentence into a true sentence in each case.

1　There are x minutes in one hour.

2　When the number x is doubled, the result is one-quarter of 48.

3　A cube has x edges.

4　x is an even prime number.

5　x months in the year have 31 days.

6　There are x possible scores in a throw of a die.

List the *solution set* of each of the sentences in questions *7-10*. For example, if z is a variable on the set $D = \{1, 2, 3, 4\}$, then the open sentence 'z is an even number' has solution set {2, 4}.

7　Here x is a variable on the set $A = \{3, 6, 9, 12, 15\}$.

a　x is an odd number　　　　　b　x is divisible by 6
c　x is a multiple of 3　　　　　d　x is less than 8
e　$x+10 = 25$　　　　　　　　　f　$x+6$ is in A

8　Here y is a variable on the set $B = \{4, 7, 10, 13\}$.

a　y is an even number　　　　　b　y is a prime number
c　y is divisible by 5　　　　　　d　y is greater than 20
e　$y+4$ is less than 14　　　　　f　$y+y = 2 \times y$

9   Here p, q, r, s are variables on the set $C = \{1, 3, 5, 7, 9\}$.

   a $p+5 = 8$ *b* q is greater than 6

   c r is less than 6   *d* $s+2 \in C$

   e $p \times p$ is in C *f* $q-8 = 0$

10   Here n is a variable on the set $E = \{1, 2, 3, 4, ..., 10\}$.

   a $n+2 = 10$ *b* $5+n$ is greater than 10

   c $n+n+n \in E$ *d* $n+3$ is less than 4

   e n is a square number   *f* $n+3 = 11-n$

---

4 Equations

An open sentence containing the verb 'is equal to' (written $=$) is called an *equation*.

Consider the equation $x+2 = 5$, where x is a variable on the set of natural numbers. Choosing replacements from this set as in Section 3, we see the following:

If $x = 1$, the sentence $1+2 = 5$ is obtained, and is false.

If $x = 2$, the sentence $2+2 = 5$ is obtained, and is false.

If $x = 3$, the sentence $3+2 = 5$ is obtained, and is true.

If $x = 4$, the sentence $4+2 = 5$ is obtained, and is false.

It is now clear that *only* when x is replaced by 3 is a true sentence obtained.

Hence the *solution* of the equation $x+2 = 5$ is $x = 3$, and the *solution set* is $\{3\}$.

We could also *solve the equation* by thinking out the answer to the question 'What number added to 2 gives 5?' The answer is 3.

Example 1.—Solve the equation $x-5 = 2\frac{1}{2}$, where x is a variable on the set of fractions.

Only when x is replaced by $7\frac{1}{2}$ do we get a true sentence ($7\frac{1}{2} - 5 = 2\frac{1}{2}$). So the solution set of the equation is $\{7\frac{1}{2}\}$.

Example 2.—Solve the equation $3+x = 1$, where x is a variable on the set of natural numbers.

There is *no* natural number which can replace x so that a true sentence is obtained. The solution set is therefore ø or $\{\ \ \}$.

Exercise 4A

Solve the following equations, assuming that the variables are on the set of natural numbers. Give the solution in the form '$x = 7$'.

1 $x + 2 = 9$	*2* $m + 1 = 7$	*3* $y + 2 = 12$
4 $p - 6 = 3$	*5* $x + 4 = 21$	*6* $4 = 9 - a$
7 $21 = z + 11$	*8* $8 = a - 3$	*9* $x + 7 = 16$
10 $x + 10 = 0$	*11* $x + 2 = 20$	*12* $16 - x = 12$
13 $m + 5 = 54$	*14* $y = 5 \times 70$	*15* $p + 1 = 0$
16 $x + 2 = 29$	*17* $x - 4 = 14$	*18* $y - 27 = 27$
19 $3 \times m = 15$	*20* $12 \times p = 96$	*21* $4 \times p = 4$
22 $4 \times n = 44$	*23* $x \times x = 49$	*24* $2 \times x = 1$
25 $\dfrac{12}{x} = 4$	*26* $\dfrac{20}{y} = 2$	*27* $\dfrac{100}{z} = 4$

Exercise 4B

Find the *solution set* of each of the following equations, the variables being defined on the set of common fractions. Give the solution set in the form $\{1\tfrac{1}{2}\}$.

1 $x + 1 = 2\tfrac{1}{2}$	*2* $y + 4 = 4\tfrac{1}{4}$	*3* $z + 4 = 7\tfrac{1}{2}$
4 $p - 1 = \tfrac{1}{2}$	*5* $q - 2 = 1\tfrac{1}{2}$	*6* $y - 3 = 3\tfrac{1}{4}$
7 $t + \tfrac{1}{4} = 5\tfrac{3}{4}$	*8* $m + \tfrac{3}{4} = \tfrac{1}{2}$	*9* $m - \tfrac{1}{4} = \tfrac{1}{2}$
10 $12\tfrac{1}{2} = w - \tfrac{1}{2}$	*11* $w + \tfrac{1}{2} = 0$	*12* $w + \tfrac{1}{2} = 1$
13 $x + x = \tfrac{1}{2}$	*14* $2 \times x = \tfrac{1}{2}$	*15* $y \times \tfrac{1}{3} = 1$
16 $q \times \tfrac{4}{3} = 1$	*17* $x \times 2 = \tfrac{1}{4}$	*18* $\tfrac{1}{2}d = 3$
19 $\dfrac{100}{v} = 10$	*20* $\dfrac{1}{100}v = 10$	*21* $\tfrac{1}{2}y = \tfrac{1}{4}$
22 $\dfrac{3}{x} = \tfrac{1}{3}$	*23* $\dfrac{3}{x} = 1\tfrac{1}{2}$	*24* $\tfrac{1}{2}t = 8\tfrac{1}{2}$

5 'Clock' arithmetic

Figure 1 shows a special kind of clock face. The pointer turns clock-wise as shown. Let us make up a new kind of arithmetic to suit this clock.

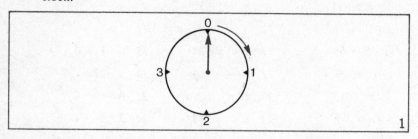

What would the time be two 'hours' after '3 o'clock'? To find out, we imagine the pointer to be at 3 and then to move on two 'hours', when it would show '1 o'clock'. Calling this operation 'addition', we will write $3+2 = 1$.

Exercise 5

1 a Copy and complete this addition table for 'clock' arithmetic using the clock face to help you.

b Does clock addition show the commutative property? Give a reason.

c Can you discover any other interesting facts from the table?

		Second number			
+		0	1	2	3
First number	0	.	.	2	.
	1	.	.	.	.
	2	2	.	.	.
	3	.	.	1	.

2 If x is a variable on the set {0, 1, 2, 3} as used in question **1** above, use the addition table you have made to solve the following equations:

  a $x+2 = 0$ **b** $x+1 = 0$ **c** $x+3 = 0$
  d $x+1 = 2$ **e** $1+x = 3$ **f** $1+x = 0$
  g $2 = x+3$ **h** $x+0 = 3$ **i** $x+1 = 3$

3 Suppose that Figure 1 represents the OFF–1–2–3 heat positions on an electric-heater rotary switch, which can turn either clockwise or anti-clockwise.

a If it is set at position 1, and is then switched off, in what two ways could this be done?

b If it is off, and then switched to 2, in what two ways could this be done?

c If it is off, and then moved five places, at which numbers might the pointer stop?

d If it is off, and then moved eight places, where would the pointer stop?

4 Figure 2 shows a six-hour clock. Starting at 2 and moving clockwise 3 hours takes the pointer to 5. Denote this operation by $2+3 = 5$. Make an addition table for the six-hour clock, as in question *1*. If x is a variable on the set $\{0, 1, 2, 3, 4, 5\}$, solve the following equations:

a $x+3 = 1$	*b* $x+4 = 0$	*c* $5+x = 1$
d $3+x = 0$	*e* $1+x = 5$	*f* $4+x = 2$

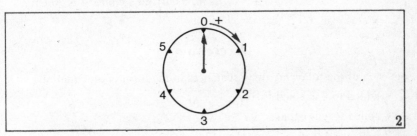

5 In Figure 2, starting at 2 and moving *anti-clockwise* 3 hours takes the pointer to 5. This operation is the inverse operation of addition and we shall call it *subtraction*. We can write $2-3 = 5$. Investigate how to get this result from the addition table.

Now try to use your table of addition to do the following subtractions:

a $3-2$	*b* $4-5$	*c* $0-2$
d $4-2$	*e* $0-5$	*f* $1-3$
g $3-1$	*h* $2-3$	*i* $2-5$

In what way does this kind of subtraction differ from subtraction of whole numbers?

6 Using your addition table for the 'six-hour' clock, solve the following equations:

a $4-x = 3$	*b* $2-x = 5$	*c* $1-x = 2$

$$d \quad x-2 = 2 \qquad e \quad x-5 = 3 \qquad f \quad x-1 = 0$$
$$g \quad x-2 = 4 \qquad h \quad x-3 = 0 \qquad i \quad 4-x = 0$$

6 Writing mathematical sentences

The first step in solving a practical problem is often to translate the information into mathematical language. This involves writing mathematical sentences about numbers, and it is useful to illustrate these sentences with diagrams (as shown in Section 7).

Example 1.—Information. '4 subtracted from x gives 7.'
         Translation. '$x-4 = 7$'.

Example 2.—Information. 'The perimeter of a triangle with sides
                         a cm, b cm and c cm long is 20 cm'.
         Translation. '$a+b+c = 20$'.

Exercise 6A

Write each of the following open sentences in the form of an equation.

1   x added to 4 is equal to 11.

2   4 added to y is equal to 9.

3   7 subtracted from x is equal to 12.

4   11 subtracted from t gives 2.

5   If 5 is added to x the answer is 14.

6   3 less than x is 8.

7   x added to 4 gives 10.

8   When 4 is subtracted from y the answer is 2.

9   When 4 is subtracted from m the answer is 6.

10   q added to q gives 14.

11   x minus y is equal to zero.

12   A number n is 4 more than 15.

13   If n is an even number, the next consecutive even number m is 2 more than n.

14   The sum of x and y is 12.

15   The difference between p and q is 29, where p is greater than q.

16   The length a cm of a rectangle is 5 cm more than the breadth b cm.

17   Adding 35 to the sum of a and b gives 360.

18   The sum of x, y and z is 180.

19   Solve the equations you obtain in questions 1–10.

Exercise 6B

Using algebraic notation, write each of the following in the form of an equation, and solve the equation.

1   30 pupils are on the roll of a class. x are absent, and the number present is 25. Find x.

2   There are 21 passengers on a bus. At the next stop x people board the bus. The number of passengers is now 30. Find x.

3   A regiment consists of n men and 100 recruits join up. The total strength is now 850. Find n.

4   A book contains 325 pages. After p pages have been read, the number still to be read is 84. Find p.

5   A farmer has 250 sheep and buys x more sheep at the market. He now has 320 sheep. Find x.

6   a and $b+5$ represent the same number. If $b = 45$, find a.

7   p and $q-20$ represent the same number. If $p = 80$, find q.

8    $r+5$ added to 15 gives 80. Find r.

9   Tom weighs 5 kg more than Jack and Jack weighs x kg. If Tom weighs 63 kg, what does Jack weigh?

10   The sum of x and y is 100. If y is 45, what is x?

11   q is equal to the sum of 4 and x.
    a   Given $x = 10$, find q.
    b   Given $q = 36$, find x.

12   x is equal to the sum of y and z.
    a   Given $y = 12$ and $z = 9$, find x.
    b   Given $x = 15$ and $y = 7$, find z.
    c   Given $x = 31$ and $z = 19$, find y.

7 Graphs of the solution sets of mathematical sentences

A ship sails 2 km east from one marker buoy to another, and then continues to sail east for a further x km, as shown.

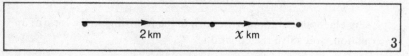

2 km x km

3

We can write down a mathematical sentence for the total distance, y km, that the ship has sailed:

$$y = 2 + x$$

If x is replaced by 1, $y = 2 + 1 = 3$

If x is replaced by 2, $y = 2 + 2 = 4$

We can now complete this table:

x	1	2	3	4	5	6
y	3	4	5	6	7	8

We can show the solution set of $y = 2 + x$ in the form of a graph obtained by plotting the points (1,3) (2, 4), (3, 5), (4, 6), (5, 7) and (6, 8) from the table, as shown in Figure 4. What do you notice about the 'dots' on the graph?

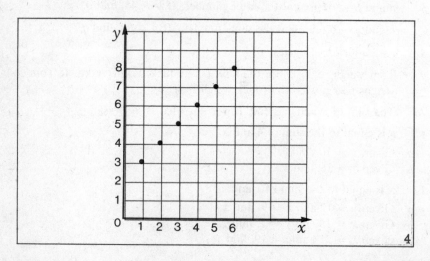

4

Exercise 7

1   Repeat the whole of the worked example above when the distances
    sailed are 1 km, then x km.

2   x and y are variables on the set of whole numbers. The relation
    between x and y is such that y is always 1 less than x.

a   From this information, write down an equation in x and y.
b   Given that x is a variable on $\{1, 2, 3, 4, 5, 6\}$, find y when x is re-
    placed by $x = 1, 2, 3, 4, 5, 6$.
c   Copy and complete this table:

x	1	2	3	4	5	6
y	0	1				

d   Show the solution set in the form of a graph, using 5-mm squared
    paper. (Take the x-axis horizontal and the y-axis vertical.)

3   Repeat the whole of question 2 when the relation between x and y is
    such that y is always double x.

4 a Draw a rectangle with length x cm and breadth 5 cm.
  b Write down a mathematical sentence for the sum y of the length and
     breadth.
  c Copy and complete this table.

x	1	2	3	4	5	6
y						

  d Show the solution set in the form of a graph, using 5-mm squared
     paper. (Take the x-axis horizontal and the y-axis vertical.)

5   Repeat the whole of question 4 for a rectangle with length x cm and
    breadth 8 cm.

6   A motor boat uses 1 litre of fuel every 4 km it sails. One day the
    boat made 3 journeys and used 5 litres of fuel.

a   What was the total length of the 3 journeys?
b   If the first trip was 10 km, the second x km and the third y km, write
    down a mathematical sentence containing x and y.
c   Now write this sentence in the form $x + y = \ldots$
d   Copy and complete this table:

x	1	2	3	4	5	6
y						

e   Show the solution set in the form of a graph on 5-mm squared paper.

Summary

1 A *sentence* may be true or false.

2 An *open sentence* is a sentence containing one or more variables, e.g. '12 is divisible by x'; '$x+y = 5$'.

3 A *variable* is a symbol which can be replaced by members of a given set.

4 The *solution set* of an open sentence is the set of replacements of the variable which give a true sentence.

   Each member of the solution set is a *solution* of the open sentence.

5 An *equation* is an open sentence containing the verb 'is equal to'.

6 The solution set of a mathematical sentence can be represented by a *graph*.

Multiplication

1 The meaning of multiplication

The product 5×3 or '5 times 3' may be taken to mean the addition of five threes, i.e.

$$5 \times 3 = \text{five threes} = 3+3+3+3+3 = 15.$$

In the same way,

$$3 \times 5 = \text{three fives} = 5+5+5 = 15.$$

As expected, $5 \times 3 = 3 \times 5$. This illustrates the commutative property of multiplication.

Similarly, $\quad 5 \times x = \text{five } x\text{s} = x+x+x+x+x,$

and $\quad\quad\quad x \times 5 = x \text{ fives} = 5+5+5+\ldots \text{ to } x \text{ terms.}$

Notice that multiplication is a kind of shorthand addition.

Exercise 1

Write each of the following in short form (e.g. $4 + 4 + 4 = 3 \times 4$):

1	$2+2+2$	*2*	$3+3$	*3*	$8+8+8+8$
4	$7+7+7+7+7$	*5*	$1+1+1+1$	*6*	$9+9$
7	$3+3+3$	*8*	$a+a$	*9*	$n+n+n$
10	$y+y+y+y+y$	*11*	$k+k+k+k$	*12*	$c+c$

13 $p+p+p+\ldots$ to 10 terms *14* $r+r+r+\ldots$ to 15 terms

15 $x+x+x+\ldots$ to 12 terms *16* $b+b+b+\ldots$ to 6 terms

Write down the meaning of each of the following
(e.g. $3 \times 7 = 7+7+7$):

17	2×5	*18*	3×8	*19*	7×2
20	5×4	*21*	6×0	*22*	4×9
23	$5 \times w$	*24*	$3 \times y$	*25*	$4 \times n$
26	$1 \times x$	*27*	$1 \times y$	*28*	$10 \times z$

2 Using the commutative law $a \times b = b \times a$

Here is a multiplication table for the set of numbers
{0, 1, 2, 3, 4, 5, 6}

				Second number			
×	0	1	2	3	4	5	6
0	0	0	0	0	0	0	0
1	0	1	2	3	4	5	6
2	0	2	4	6	8	10	12
First number 3	0	3	6	9	12	15	18
4	0	4	8	12	16	20	24
5	0	5	10	15	20	25	30
6	0	6	12	18	24	30	36

Exercise 2

By looking along a *row* of the table, find the solution of each of these equations:

1 $4 \times x = 24$ *2* $2 \times x = 12$ *3* $5 \times x = 15$

4 $3 \times x = 15$ *5* $3 \times x = 3$ *6* $5 \times x = 0$

7 $3 \times x = 6$ *8* $1 \times x = 5$ *9* $4 \times x = 20$

By looking down a *column* of the table, solve these equations:

10 $x \times 4 = 24$ *11* $x \times 2 = 8$ *12* $x \times 5 = 15$

13 $x \times 3 = 15$ *14* $x \times 3 = 3$ *15* $x \times 3 = 12$

16 $x \times 3 = 6$ *17* $x \times 4 = 20$ *18* $x \times 1 = 3$

Note that the answers to questions *1* and *10* above are the same because of the commutative law of multiplication, which states that in the multiplication of two numbers the order in which the numbers are taken does not matter.

What other pairs of questions have equal answers for that reason?
It is usual to write $3 \times x$ as $3x$, the multiplication sign being

omitted. In the same way, $x \times 3$ can be written $x3$. But as $3 \times x = x \times 3$ (commutative law of multiplication), $x3$ is always written $3x$, i.e. the numerical factor is placed before the variable. The numerical factor is called a *coefficient*. Thus, the coefficient of x in $3x$ is 3.

For $x \times x$, we write x^2 ('x squared') in preference to xx.

Geometrical illustrations.—In each of the diagrams, the unit of length is the centimetre so that area will be measured in square centimetres (cm²).

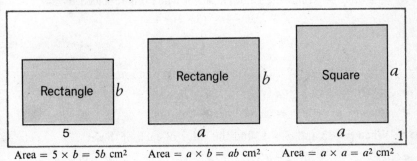

Area $= 5 \times b = 5b$ cm²   Area $= a \times b = ab$ cm²   Area $= a \times a = a^2$ cm²

Exercise 3A

Write in shorter form (e.g. $7 \times m = 7m$; $x \times 1 = 1x = x$):

1	$3 \times m$	*2*	$7 \times p$	*3*	$2 \times x$	*4*	$12 \times p$
5	$1 \times y$	*6*	$8 \times c$	*7*	$y \times y$	*8*	$x \times 3$
9	$r \times 9$	*10*	$t \times 10$	*11*	$h \times 12$	*12*	$k \times 15$
13	$c \times c$	*14*	$x \times y$	*15*	$p \times q$	*16*	$m \times m$

17 Write down an expression for the area of each of the following, in which the dimensions are in centimetres:

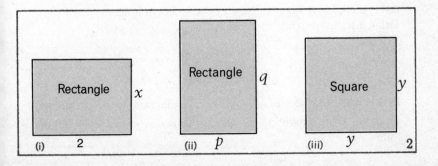

Solve the following equations, the variables being on the set of natural numbers {1, 2, 3, 4, ...}:

18 $3m = 12$ *19* $7p = 14$ *20* $2x = 18$ *21* $12p = 36$

22 $5y = 35$ *23* $8c = 96$ *24* $9n = 9$ *25* $5x = 10$

26 $y^2 = 9$ *27* $t^2 = 81$ *28* $z^2 = 1$ *29* $14 = 2y$

30 $45 = 3v$ *31* $3+w = 45$ *32* $14 = 2+y$ *33* $5h = 5$

Example.—Find the value of $3n-4$ when n is replaced by 2.
$$3n-4 = (3 \times 2)-4$$
$$= 6-4$$
$$= 2$$

34 Find the value of each of the following when $n = 2$:

a $n+6$ *b* $2n$ *c* $3n+1$ *d* $5n-1$
e $10n-5$ *f* n^2 *g* $10-4n$ *h* $20-8n$

35 When $p = 2$ and $q = 1$, find the value of each of these:

a $p+q$ *b* $p-q$ *c* pq *d* $3p+2q$
e $2p-2q$ *f* p^2+q^2 *g* $5p+3q$ *h* p^2-q^2

36 When $a = 0$, $b = 1$, $c = 3$, find the value of each of these:

a $a+b+c$ *b* abc *c* $a+2b+3c$ *d* $5a+5b+5c$
e $a^2+b^2+c^2$ *f* $3abc$ *g* $11b-2c$ *h* $c^2-b^2-a^2$

37 Here are two straight lines, AB and PQ. The unit of length is the centimetre.

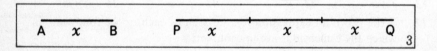

AB is x cm long. PQ is $(x+x+x)$ cm $= 3x$ cm long.
Draw lines to illustrate in the same way *a* $2x$ *b* $6x$.

Exercise 3B

In this Exercise, the unit of length for the diagrams is the centimetre.

1 Find the total length of the sides (the *perimeter*) of each of the following:

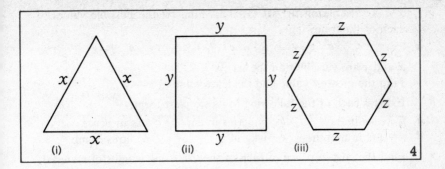

(i)   (ii)   (iii)   4

2   Figure 5 shows a skeleton cube, made of wire, whose sides are *t* cm long.
a   Write down the total length of wire in the cube.
b   Write down an expression for the area of one face of the cube. Hence give the total area of all the faces in terms of *t*.
c   Calculate the total length of wire in *a* when *t* is replaced by 10.

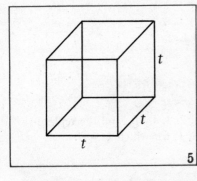

5

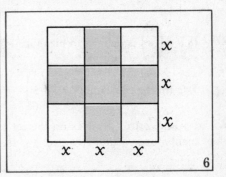

6

3   Figure 6 shows a cross drawn in a square.
a   Write down the perimeter of the square.
b   Write down the perimeter of the cross.
c   Write down an expression for the area of the cross in terms of x.
d   Calculate the perimeter of the cross in *b* when $x = 2$.

4   Find the value of each of the following when *n* is replaced by 5:
a $n+2$ *b* $4n$ *c* n^2 *d* $2n+1$
e $5n-5$ *f* n^2-25 *g* $10-2n$ *h* n^2+2n+1

5   When $p = 3$ and $q = 2$, find the value of each of these:
a $p-q$ *b* $2p+3q$ *c* p^2+q^2 *d* $6p-9q$

6 x is a variable on the set $\{1, 2, 3\}$. Find all the possible values of each of the following:

   a $2x$ *b* $x+8$ *c* $3x-3$ *d* $9-x^2$

7 x and y are variables on the set $\{0, 1, 2, 3\}$.
Find the greatest value and the least value of x^2+y^2.

8 Express each of the following in its simplest form:

   a k weeks in days   *b* p metres in cm   *c* £q in pence
   d x years in months   *e* y kg in g   *f* z litres in ml

9 Find the cost in pence of each of the following lengths of material:

   a 5 metres at 50 pence per metre   *b* 5 metres at x pence per metre
   c x metres at 50 pence per metre   *d* x metres at x pence per metre

10 Copy and complete the following multiplication tables:

a

×	x	y	z
0	0	.	.
1	.	y	.
2	.	.	$2z$

b

×	i	j	k
i	i^2	ij	.
j	.	.	jk
k	.	.	.

11 $(a\,b\,c)\begin{pmatrix} x \\ y \\ z \end{pmatrix}$ is defined to mean $ax+by+cz$.

Use this to calculate $(1\ 2\ 3)\begin{pmatrix} 4 \\ 5 \\ 6 \end{pmatrix}$

12 x, y and z are variables on the set $\{1, 2, 3, 4, 5\}$. Find the replacements for x, y and z, with x less than y, for which $x^2+y^2 = z^2$.

---

3 Using the associative law

$$(a \times b) \times c = a \times (b \times c) = abc$$

We know that $3 \times 4 = 12$. How do we interpret $3 \times 4 \times 5$?
Replacing 3×4 by 12, $3 \times 4 \times 5 = 12 \times 5 = 60$.
Replacing 4×5 by 20, $3 \times 4 \times 5 = 3 \times 20 = 60$.
Hence $3 \times 4 \times 5 = (3 \times 4) \times 5 = 3 \times (4 \times 5)$.

   This example illustrates the associative law of multiplication, which states that the way in which factors are grouped does not matter.

Using also the commutative law of multiplication,

$$3 \times 4 \times 5 = (3 \times 4) \times 5 = (4 \times 3) \times 5 = 5 \times (4 \times 3) = 5 \times 4 \times 3$$

and $3 \times 4 \times 5 = 3 \times (4 \times 5) = 3 \times (5 \times 4) = (5 \times 4) \times 3 = 5 \times 4 \times 3$

so that the order of factors does not matter.

These statements are true for any number of factors, so there is no need to put in brackets unless we wish to show how a product is obtained.

Example 1. $3 \times a \times 5 \times b = 3 \times 5 \times a \times b$ (picking out coefficients first)

$$= (3 \times 5) \times (a \times b)$$
$$= 15 \times ab$$
$$= 15ab$$

Example 2. $2a \times 3b \times 4c = (2 \times 3 \times 4) \times (a \times b \times c)$

$$= 24abc$$

Note that the numerical factor, or *coefficient*, is always placed first, followed by the variables in alphabetical order.

Just as $x \times x$ is written x^2 ('x squared'), so $x \times x \times x$ is x^3 ('x cubed').

Geometrical illustrations.—In each of the diagrams the unit of length is the centimetre, so that volume will be measured in cubic centimetres (cm^3).

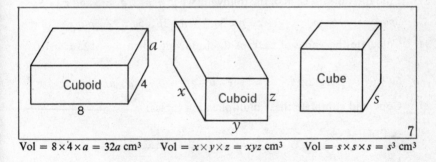

Vol $= 8 \times 4 \times a = 32a$ cm³   Vol $= x \times y \times z = xyz$ cm³   Vol $= s \times s \times s = s^3$ cm³

Exercise 4A

By changing the order, calculate each of the following in as easy a way as possible:

1	$2 \times 9 \times 5$	*2*	$5 \times 38 \times 20$	*3*	$50 \times 7 \times 2$
4	$2 \times 5 \times 19 \times 10$	*5*	$20 \times 26 \times 5$	*6*	$25 \times 15 \times 4$

Write each of the following in its simplest form (e.g. $5 \times a \times 3 = 15a$):

7	$3 \times 2 \times a$	8	$4 \times 1 \times p$	9	$8 \times 5 \times r$
10	$5 \times a \times b$	11	$3 \times m \times n$	12	$4 \times 2 \times t$
13	$2 \times x \times 3$	14	$1 \times x$	15	$g \times h \times 7$
16	$a \times a \times b$	17	$c \times e \times c$	18	$5 \times 3w$
19	$u \times 8v$	20	$2x \times 4y$	21	$5m \times 2n$

22   Write down an expression for the volume of each object in Figure 8, in which the dimensions are in centimetres:

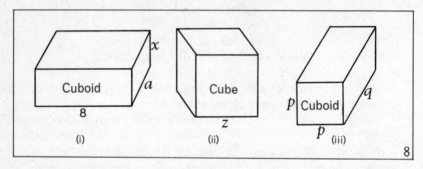

(i)   (ii)   (iii)

8

23   Find the value of each of the following when $p = 2, q = 3$ and $r = 5$:

a $3pq$ b pqr c $p \times 3q$ d $2p \times 5q$ e $p^2q^2r^2$

24   Calculate the value of each of the following when $u = 12, g = 10, t = 2$:

a $5u + 6g$ b $2gt^2$ c $u + gt$ d $ut + \frac{1}{2}gt^2$

25   Copy and complete these multiplication tables:

a

×	x	$2y$	$3z$
1	.	$2y$	.
2	$2x$	.	.
3	.	.	.

b

×	x	$2y$	$3z$
x	.	$2xy$	.
$2y$	.	.	.
$3z$	.	.	$9z^2$

26   If $x = 2, y = 3, z = 5$, calculate the values of:

a $(x+y) \times (y+z) \times (z+x)$ b $x^2 \times y^2 \times z^2$

Exercise 4B

Write each of the following in its simplest form,
e.g. $3 \times a \times 4 \times b = 12ab$:

1   $2 \times 8 \times m \times n$ 2 $3 \times u \times 7 \times v$ 3 $2 \times 3 \times x \times x$

4   $3 \times g \times g \times h$ 5 $a \times a \times b \times c$ 6 $a \times a \times b \times b$

7   $2k \times 3 \times h$ 8 $2 \times 4y \times 5x$ 9 $a \times 3b \times 3c$

10   $2 \times 5x \times 9y$ 11 $2a \times 3b \times 4c$ 12 $3p \times q \times 2p \times 0$

13   When $p = 1, q = 3, r = 5$, find the value of each of these:

   a $p+q+r$ b $q+r-8p$ c pqr
   d $10p+10q+10r$ e $pq+qr+rp$ f $pq+3qr+5rp$

14   When $u = 1, v = 4$ and $w = 5$, find the value of each of these:

   a $3uv+2v+1$ b $u^2+v^2+w^2$ c $2uv+2vw$
   d $u^3+v^3+w^3$ e $uvw+vw+w$ f $(u+v+w)^2$

15   When $x = 2$ and $y = 3$, calculate the values of:

   a $x^2+3xy+1$ b $3x^2+2xy+y^2$ c $10x^2y^3$

16   Let us make up a multiplication table for our 'clock' arithmetic.

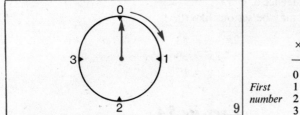

×	0	1	2	3
0	.	.	.	.
1	.	.	.	.
2	.	.	.	2
3	.	.	.	.

Second number across the top, *First number* down the side.

We know that $2 \times 3 = 3+3$, so to find the value of 2×3 we follow
the pointer round to 3 and then add on 3, to finish at 2. So,
$2 \times 3 = 2$ and we enter this answer in the table.

a   Now copy the table and enter all the answers.

b   Is the table symmetrical about the main diagonal? Can you see
anything else of interest about the table?

c   If x is a variable on the set $\{0, 1, 2, 3\}$ above, use the multiplication
table you have made to find the solution sets of:

(1) $2x = 2$ (2) $3x = 0$ (3) $2 = 3x$

(4) $3x = 1$ (5) $2x = 0$ (6) $2x = 1$

4 Equations

Example 1.—Solve the equation $2x+7 = 33$, where x is a variable on the set of natural numbers.

To solve this equation we think, 'what number added to 7 gives 33?' The answer is 26.

Thus, if we want to make $2x+7 = 33$ a true sentence, $2x$ must be equal to 26, and so $x = 13$.

The working may be set out like this:

$$2x+7 = 33$$
$$2x = 26$$
$$x = 13$$

Example 2.—Solve the equation $2x-6 = 33$, where x is a variable on the set of fractions.

We think, 'what number gives 33 when 6 is subtracted from it?' The number is 39. So to make $2x-6 = 33$ a true sentence, $2x$ must be equal to 39. Hence $x = 19\frac{1}{2}$.

The working may be set out like this:

$$2x-6 = 33$$
$$2x = 39$$
$$x = 19\frac{1}{2}$$

Exercise 5A

State whether each of the following is true (T) or false (F):

1 $a+5 = 10$ when $a = 5$ 2 $b+2 = 11$ when $b = 9$

3 $2c+1 = 7$ when $c = 3$ 4 $2d-1 = 5$ when $d = 2$

5 $e+3 = 3$ when $e = 0$ 6 $5f+5 = 40$ when $f = 9$

Solve the following equations, where x, y, z, etc., are variables on the set of whole numbers $\{0, 1, 2, 3, ...\}$:

7 $2x+1 = 5$ 8 $3y+3 = 9$ 9 $4z+1 = 21$

10 $2p-1 = 7$ 11 $5q+5 = 55$ 12 $6r-1 = 17$

13 $7a+7 = 7$ 14 $7a+7 = 14$ 15 $7a-2 = 12$

16    $2x+3 = 5$ 17 $2x+3 = 7$ 18 $2x+3 = 3$

19    $8y+6 = 54$ 20 $9z+4 = 40$ 21 $10p-10 = 90$

22    $10-x = 7$ 23 $10-2x = 2$ 24 $10-2x = 0$

Solve the following equations, where x is a variable on the set of fractions:

25    $2x+1 = 6$ 26 $2x-1 = 2$ 27 $x+\frac{1}{2} = 8\frac{1}{2}$

28    $x+\frac{1}{2} = 5$ 29 $3x+1 = 2$ 30 $3x-1 = 1$

Exercise 5B

State whether each of the following is true (T) or false (F):

1    $a+4 = 12$ when $a = 8$ 2 $b+4 = 4$ when $b = 0$

3    $2c+8 = 12$ when $c = 3$ 4 $3d-1 = 10$ when $d = 3$

5    $2p = p+3$ when $p=3$ 6 $5q+4 = 3q+12$ when $q = 4$.

Solve the following equations, where x, y, z, etc., are variables on the set of whole numbers $\{0, 1, 2, 3, ...\}$:

7    $2x+1 = 7$ 8 $5z+5 = 35$ 9 $6x+1 = 25$

10    $3x+2 = 5$ 11 $3p+4 = 16$ 12 $2x+15 = 31$

13    $2x-1 = 9$ 14 $4x-5 = 19$ 15 $7y-4 = 17$

16    $8x-11 = 29$ 17 $12-2t = 4$ 18 $25 = 6x+7$

19    $3x+4 = 4$ 20 $47-6y = 5$ 21 $29 = 2x - 9$

22    $17p = 306$ 23 $101 = 4m+1$ 24 $13t+142 = 259$

Solve the following equations, where x is a variable on the set of fractions:

25    $2x+6 = 11$ 26 $2x-3 = 12$ 27 $34-2x = 7$

28    $\frac{1}{2}x = 12$ 29 $\frac{1}{2}x+1 = 3$ 30 $\frac{2}{3}x-1 = 3$

31   x is a variable on the set $\{0, 1, 2, 3, 4, 5\}$. Find the solution set of each of the following equations:

   a $x^2 = 4x$ b $x^2+2x = 0$ c $(x+1)^2 = 4$ d $x^2+5 = 6x$

32   x and y are variables on the set $\{1, 2, 3\}$. Can you find the *two* pairs of replacements for x and y such that $x+2y = 7$?

5 Using the distributive law $ab+ac = a(b+c)$

The distributive law enables us to simplify algebraic expressions.

Example 1. $3x+4x = (3+4)x = 7x$ (notice the common factor x)

Example 2. $10a-9a = (10-9)a = 1a = a$ (notice the common factor a)

Example 3. $4p+8p-3p = (4+8-3)p = 9p$ (notice the common factor p)

But $3x+4y$ cannot be simplified (there is no common factor).

Exercise 6A

Simplify the following, noting the common factor in each case:

1	$2x+3x$	2	$2k+5k$	3	$4a+6a$
4	$4m+4m$	5	$3h+8h$	6	$8u+7u$
7	$2c+8c$	8	$9t+t$	9	$y+5y$
10	$10b+3b$	11	$5a-4a$	12	$12c-11c$
13	$5p-5p$	14	$8n-n$	15	$5x^2+3x^2$
16	$14a^2-13a^2$	17	$3ab+10ab$	18	x^3+4x^3

19 Simplify each of the following, where possible:

    a $4w+2w$ b $4g+7h$ c $10x-3x$

    d $5x+5x$ e $8a-3b$ f $9a+2a$

Solve the following equations, where x, y, z, etc., are variables on the set of whole numbers:

20 $2x+3x = 40$ 21 $8y-5y = 24$ 22 $7z+7z = 14$

23   $9p-5p = 4$ 24 $6q+12q = 54$ 25 $5r+5r = 0$

Simplify the following:

26   $2x+3x+4x$ 27 $2p+5p+p$ 28 $9m+7m+4m$

29   $2h+2h+2h$ 30 $6w+w+2w$ 31 $7y-4y+8y$

32   $12x+2x-6x$ 33 $5y+3y-8y$ 34 $6z+11z-16z$

35   $x^2+3x^2+8x^2$ 36 $2y^2-y^2+y^2$ 37 $z^2+3z^2+5z^2$

The expression $3x+6y+4x-5y$ contains four *terms*. These are $3x$, $6y$, $4x$ and $5y$. $3x$ and $4x$ are called *like* terms; $6y$ and $5y$ are also like terms. Such expressions can be simplified by means of the distributive law; in effect, we add or subtract the coefficients of like terms.

Example. $\begin{aligned} 3x+6y+4x-5y &= (3x+4x)+(6y-5y) \\ &= (3+4)x+(6-5)y \\ &= 7x+y \end{aligned}$

Exercise 6B

Simplify each of the following, where possible:

1   $8x+7x$ 2 $12y-7y$ 3 $6a+5a-3a$

4   $10b-7b+2b$ 5 $8c-6c-2c$ 6 $5x+6y$

7   $5x+3x+6y$ 8 $5x+6y-2y$ 9 $5x+3x+6y-2y$

10 $4a+9a-6a$ 11 $3a+5a+7b-2b$ 12 $6c+2d+3c-d$

13 $2x^2+4y^2+3x^2-2y^2$ 14 $x^2+y^2+8x^2$ 15 $a^2+b^2+2a^2-b^2$

16   Find a simple expression for the total length of each of the following lines, the unit of length being the centimetre:

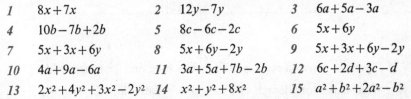

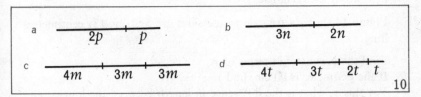

17   From Figure 11(i), write down the length of AC in terms of a. If AC = 40 cm, find a.

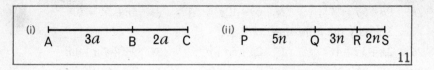

18 From Figure 11(ii), write down the length of PS in terms of n. If PS = 95 cm, find n.

19 Figure 12 shows a square of side $2p$ cm.

a Write down the length of the perimeter in terms of p.
b If $p = 5$, what is the perimeter?
c If the perimeter measures 32 cm, find p.

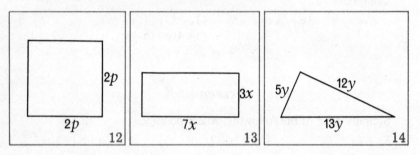

20 Figure 13 shows a rectangle $7x$ metres long and $3x$ metres wide.

a What is the total length of a long and a short side?
b What is the perimeter?
c If the perimeter is 100 metres, find x and hence state the length and breadth of the rectangle.

21 Figure 14 shows a triangle whose sides are $5y$, $12y$, $13y$ centimetres long.

a Find the perimeter in terms of y.
b If the perimeter is 90 cm, find y.
c For this y, what is the difference in length between the longest and shortest side?

22 In this question, the metre is the unit of length. Find the perimeter of each of the following figures in its simplest form.

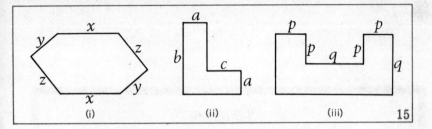

(i)   (ii)   (iii)   15

23   A rectangular carpet is x metres long and 3 metres wide.

 a   Make a sketch of the carpet and mark in the dimensions.
 b   Write down the perimeter in terms of x, giving the result in its simplest form.
 c   If the perimeter is 16 metres, form an equation in x.
 d   Hence find x and state the area of the carpet.

24   A boy ran $3d$ metres and then swam $7d$ metres.

 a   What total distance did he travel?
 b   If he covered a total distance of 80 metres, form an equation in d and solve it.
 c   What distance did he swim for this value of d?

25   Betty scores 15 marks more than Anne in a mathematics test.

 a   Suppose that Anne's mark is x, what is Betty's mark in terms of x?
 b   Hence find their combined marks total.
 c   Their combined marks total 137. Form an equation in x. Solve it and hence state Betty's mark.

Summary

1   *Multiplication is commutative*
    e.g. $3 \times a = a \times 3$, and is written $3a$.

2   *Multiplication is associative*
    e.g. $(3 \times a) \times b = 3 \times (a \times b) = 3ab$.

3   *Multiplication is distributive over addition*
    e.g. $3a + 5a = (3 + 5)a = 8a$.

4   *The coefficient in a term is the numerical factor in the term*
    e.g. the coefficient of $3a$ is 3.

5   *The value of $3x^2$ when $x = 2$ is $3 \times 2 \times 2 = 12$.*

6   $a \times a = a^2$, and $a \times a \times a = a^3$.
    Also $a + a = 2a$, $a + a + a = 3a$.

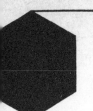

Revision Exercises

Revision Exercises on Chapter 1
An Introduction to Sets

Revision Exercise 1A

1 Using the { } brackets notation, list each of the following sets:

  a The set of months ending in the letter *y*.
  b The set of the first ten prime numbers.
  c The set of letters of the alphabet between K and R.
  d The set of odd numbers greater than 20 but less than 30.

2 Describe the following sets in words:

  a {January, July, June}
  b {Mercury, Venus, Earth, Mars, Jupiter, Saturn, Uranus, Neptune, Pluto}
  c {cone, pyramid}   *d* {2, 3, 32, 23}

3 Connect these elements with their sets, using '∈':
    Element: a, green, 12, b, 13.
    Sets: $\{a, b, c\}$, {even numbers}, {colours}, {10, 11, 12, 13}, {vowels in English alphabet}, {red, yellow, green}.

4 State which of the following are true and which are false:

  a $2 \in$ {prime numbers}.
  b $\{x, y, z\}$ and $\{p, q, r\}$ are equal sets.
  c {0} is the empty set.
  d $\{k, l, m, n\} = \{m, l, k, n\}$
  e If $A =$ {whole numbers greater than 50}, then $46 \notin A$.

5 Using set notation, rewrite the following:

  a 3 is a member of set W.
  b The empty set.
  c x does not belong to A.
  d S is a subset of T.
  e The set P is equal to the set Q.

6 $S = \{1, 2, 3, 4, 5, 6, 7, 8, 9, 10\}$. List the following subsets of S:

a The set of prime numbers in S.

b The set of odd numbers in S less than 10.

c The set of elements in S which are factors of 70.

d The set of numbers in S which are divisible by 3.

7 $E = \{1, 2, 3, 4, 5\}$.

a List all the possible subsets of E that contain three elements.

b How many subsets are there?

c Investigate whether or not the intersection of any two of these subsets is the empty set.

8 Here are four sentences and four Venn diagrams:

a $A \subset B$ **b** $A = B$ **c** $B \subset A$ **d** $A \cap B = \emptyset$

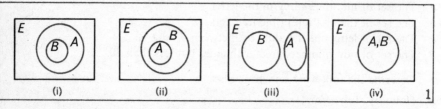

(i)   (ii)   (iii)   (iv)

Pair off each sentence with a Venn diagram.

Copy the diagrams and shade in, where you can, the part which shows $A \cap B$.

9 Find a set equal to each of the following:

a $\{1, 2, 3\} \cap \{2, 3, 4, 5\}$ **b** $\{1, 2, 3\} \cap \{4, 5, 6\}$

c $\{1, 2, 3\} \cap \{3, 1, 2\}$ **d** $\emptyset \cap \{1, 2, 3\}$

10 $E = \{a, b, c, d, e, f, g\}$; $P = \{a, b, c\}$ and $Q = \{c, b, e, d\}$.

a Find $P \cap Q$.

b Draw a Venn diagram showing the relation between the sets, and indicate the elements belonging to each set.

c Which elements of E do *not* belong to P or Q?

11 $E = \{1, 2, 3, 4, 5, 6, 8, 10\}$; $A = \{1, 2, 3, 4\}$, $B = \{3, 4, 5\}$ and $C = \{2, 4, 6, 8, 10\}$.

a Find $A \cap B$, $B \cap C$ and $A \cap C$.

b The set of elements common to A, B and C is denoted by $A \cap B \cap C$. Find $A \cap B \cap C$.

c Try to construct a Venn diagram showing the sets above and their relations.

12 Given that $A = \{0, 1, 2\}$, which of the following are true?

  a $2 \in A$ *b* $1 \subset A$ *c* $\{1\} \subset A$ *d* $0 \in \emptyset$ *e* $A \subset A$

Revision Exercise 1B

1 List the following sets:

  a {Numbers on a telephone dial}

  b {Leap years between 1982 and 1990}

  c {Names of pupils in your row in class}

  d {Human senses}

2 $S = \{5, 7, 9, 11, 12, 13\}$.

  a Make a new set by adding 3 to each member of S.

  b List the set of multiples of 3 which belong to S.

  c List the set of prime numbers belonging to S.

  d List the set of different numbers obtained by subtracting each member of S from the next greater number in S.

3 $A = \{a, b, c, d\}$. List the following sets:

  a Each of its elements contains *two* letters of A, e.g. *ac*.

  b Each of its elements contains *three* letters of A.

  c Each of its elements contains *four* letters of A.

4 Which of the following are true and which are false?

  a 10101 is a member of the set of whole numbers.

  b 10101 is a member of the set of even numbers.

  c If M is the set of whole numbers which are divisible by 2 and also by 3, then $84 \in M$.

5 Use the symbol $\in$ to show of which set, or sets, each of the following objects is an element:

  Objects: 2, elephant, Paris, 3, Dublin.

  Sets: {odd numbers}, {first four whole numbers}, {capitals of Europe}, {even numbers}, {animals}

6 Which of the following pairs of sets are equal?

  a $\{b, o, y, s\}, \{g, i, r, l, s\}$ *b* $\{b, o, y, s\}, \{s, o, b, y\}$

  c {even numbers between 1 and 9}, {multiples of 2 from 2 to 8 inclusive}.

7 If $M = \{1, 2, 3, 4, 5, 6, 7, 8, 9\}$, list the following subsets of M:

  a If 1 is subtracted from each of its members the answers are divisible by 2.

b If 1 is added to each of its members the answers are *not* divisible by 2.

c Its members are factors of 21.

d Its members are factors of 20.

e Its members are the digits used in the number of days in a year.

8 List the following sets:

a $A = \{\text{prime factors of 210}\}$

b $B = \{\text{odd numbers less than 11}\}$

c $C = \{\text{common factors of 24 and 60}\}$

d $A \cap B$ *e* $B \cap C$

9 A and B are subsets of a universal set E.

Draw a Venn diagram to illustrate each of the following:

a Each member of A is a member of B.

b Some members of A are also members of B.

c No member of A is a member of B.

d Each member of A is a member of B and each member of B is also a member of A.

10 $S = \{\text{roses in the local florist's}\}$, $C = \{\text{yellow flowers in this shop}\}$.

a Describe in words $S \cap C$.

b On a certain day $S \cap C = \emptyset$. Interpret this in words.

11 Here are four pairs of sets, each contained in the universal set of whole numbers:

a The set of even numbers; the set of multiples of 4.

b The set of whole numbers less than 10; the set of all possible remainders when the members of the universal set are divided by 10.

c The set of multiples of 4; the set of numbers exactly divisible by 5.

d The set of odd numbers; the set of multiples of 4.

Which of the four diagrams given below correctly illustrates the pairing of the sets?

(i)

(ii)

(iii)

(iv)

2

12 $E = \{$natural numbers$\}$; $A = \{$odd numbers less than 20$\}$,
$B = \{$even numbers less than 20$\}$.
If $C = \{$multiples of 5 less than 20$\}$, write down $B \cap C$ and $A \cap C$.
What set is $A \cap B$?

13 A is the set of all people over 21 years of age in the British Isles;
M is the set of Members of Parliament. Draw a Venn diagram to
illustrate $A \cap M$. Would the diagram be the same for $M \cap A$?

14 $P = \{$solids which have at least one corner$\}$, $Q = \{$cuboids$\}$,
$R = \{$spheres$\}$, $S = \{$cones$\}$.
Which of the following are true?

   a $P \subset Q$ *b* $Q \subset P$ *c* $P \cap R = \emptyset$ *d* $S \subset P$
Draw Venn diagrams to illustrate each of the true statements.

15 $E = \{0\ 1, 2, 3, 4, 5, 6, 7, 8, 9, 10\}$; $A = \{1, 2, 3, 5, 7, 9\}$,
$B = \{2, 3, 5, 6, 8, 10\}$, $C = \{2, 3, 4, 7, 8, 9\}$.
List the following sets:

   a $A \cap B$ *b* $B \cap C$ *c* $A \cap C$ *d* $A \cap B \cap C$
Show these intersections in a Venn diagram.

16 X, Y and Z are subsets of a universal set E such that $X \subset Y$ and
$Y \subset Z$. Illustrate this information by a Venn diagram. What is
$X \cap Z$?

Revision Exercises on Chapter 2
Mathematical Sentences   1. Equations

Revision Exercise 2A

1   State whether each of the following sentences is true or false:

   a $14 + 19 = 33$ *b* $87 - 78 = 19$ *c* $8 \times 0 = 0$
   d   The product of two odd numbers is always an even number.
   e   The year 1996 will be a leap year.
   f   15 metres = 1500 centimetres.

2   In each of the following open sentences, state whether the sentence
obtained is true or false for the given replacement of the variable:

   a $x + 4 = 13$; $x = 9$ *b* $z + 1 = 0$; $z = 0$
   c $x - 25 = 50$; $x = 25$ *d* p is a factor of 57; $p = 3$
   e $5 \times y = 0$; $y = 0$ *f* $t \div 8 = 9$; $t = 72$

3 Find the solution set of each of the following open sentences, the variables being on the set $S = \{1, 2, 3, 4, 5, 6, 7, 8, 9, 10\}$.

a $x + 1 = 9$ b $y - 5 = 0$ c $7 - a = 3$
d $n + n = 10$ e $m + 6$ is in S f $p \times p$ is in S

4 Find the solution set of the following open sentences, the variables being on the set W of whole numbers:

a x divides 12 exactly
b z is a common factor of 24 and 30
c y is less than 5
d p is a prime number between 20 and 30
e n is an odd number between 2 and 10
f $t + t = 18$

5 Solve the following equations, assuming that the variables are on the set of natural numbers. If there is no solution, say so.

a $7 + x = 12$ b $x - 7 = 12$ c $y - 8 = 15$
d $8 - y = 15$ e $z - 36 = 0$ f $\frac{1}{4}m = 1$
g $\frac{6}{n} = 2$ h $x + x + x = 27$ i $\frac{1}{3}x = 3$

6 Given that $E = \{1, 2, 3, 4, 5, 6, 7, 8\}$, invent open sentences whose solution sets are:

a $\{3\}$ b $\{1, 2, 3\}$ c $\{2, 4, 6, 8\}$
d $\{2, 3, 5, 7\}$ e $\{6, 7, 8\}$ f $\emptyset$

7 Here is a special kind of addition table:

+	0	1	2	3
0	0	1	2	3
1	1	2	3	10
2	2	3	10	11
3	3	10	11	12

Given that x is a variable on the set of numbers in the table, solve the following equations:

a $x + 2 = 3$ b $3 + x = 11$ c $x + 2 = 2$
d $x + x = 10$ e $x - 1 = 2$ f $x - 3 = 3$

8 Write each of the following open sentences in the form of an equation, using algebraic notation:

a t added to 10 gives 24.
b x subtracted from 5 gives 1.
c y increased by 7 gives 35.   d x times x is equal to 169.

e When 8 is taken away from *a*, the result is *b*.

f The temperature increases by $t°$ from 15°. The temperature is now 21°.

Revision Exercise 2B

1 State whether each of the following sentences is true or false:

a $7897 \times 8939 = 7053612$.

b Every year contains exactly 365 days.

c If *a* is an odd number and *b* is an even number then $a+b$ is an odd number.

d $\{1, 2\} \subset \{1, 2, 3\}$

e 17 is a factor of 51

f $\{0\} \neq \varnothing$

2 In the following open sentences the variables are on the set $A = \{2, 4, 6, 8, 10\}$. Find the solution set in each case:

a $p-3 = 5$ *b* *r* is greater than 7

c $a+a \notin A$ *d* $x-1$ is an even number

e $y+1$ is not a prime number   *f* *p* divided by 2 belongs to *A*

3 Solve the following equations where the variables are on the set *N* of natural numbers:

a $x+19 = 21$ *b* $2+y = 21$ *c* $m+4 = 2m$

d $4+m = 4$ *e* $y+\frac{1}{2}y = 6$ *f* $\frac{1}{5}n = 20$

g $\frac{5}{n} = 20$ *h* $\frac{x}{x} = 1$ *i* $w \times w = 225$

4 $E = \{1, 2, 3, 4, 5, 6, 7, 8, 9, 10\}$, and *x* is a variable on *E*. Find the solution set of each of the following open sentences:

a $x+7$ is in *E*

b *x* is greater than 5 *and* less than 8

c *x* is greater than 8 *or* less than 5

d $x \times x$ is in *E*

e $x \times x \times x$ is in *E*

f $2 \times (x-9) = 0$

5 The addition table for $\{e, a, b, c\}$ is shown here, and *x* is a variable on this set. Solve the following equations:

a $a+x = b$ *b* $x+b = a$

c $x+x = e$

+	e	a	b	c
e	e	a	b	c
a	a	e	c	b
b	b	c	e	a
c	c	b	a	e

6 x is a variable on $\{1, 2, 3, ..., 9, 10\}$ and y is a variable on the set of whole numbers.

The sentence $y = x + 10$ is to become a true sentence for replacements of x from the above set.

a Find y when x is replaced by 1.

b Find y when x is replaced by 2.

c Copy and complete the following table which shows the set of pairs (x, y) such that $y = x + 10$:

x	1	2	3	4	5	6	7	8	9	10
y	11	12								

d Using 5-mm squared paper, draw a graph to illustrate the relation between y and x.

7 x is a variable on $\{1, 2, 3, 4\}$ and y is a variable on $\{7, 8, 9, 10\}$. List the set of pairs (x, y) such that:

a $x + y = 12$ *b* $x + y$ has its greatest value

c $x + y$ has its least value   *d* x times y is greater than 30

8 p and q are variables on $\{1, 2, 3, 4, 5, 6, 7, 8\}$. List the set of pairs (p, q) such that:

a $p - q = 4$

b $p + q$ is greater than $p \times q$

Revision Exercises on Chapter 3. Multiplication

Revision Exercise 3A

1 Write in shorter form:

a $14 \times k$ *b* $c \times 8$ *c* $a \times b$ *d* $a \times a$ *e* $b \times b \times b$

2 Write down expressions for the areas of these rectangles:

Length (cm)	8	4	p	y	a
Breadth (cm)	6	x	q	y	12

3 Write down expressions for the perimeters of the rectangles in question **2**.

4 Solve the following equations, the variables being on the set of natural numbers:

a $5x = 50$ *b* $7x = 84$ *c* $8y = 8$ *d* $z^2 = 36$

5 Find the value of each of the following when $n = 5$:

  a $n+12$ **b** $3n-1$ **c** n^2+7 **d** $2n^2$ **e** n^3

6 Write each of the following in its simplest form:

  a $7\times4\times a$ **b** $5\times x\times y$ **c** $2\times p\times3\times q$ **d** $4\times k\times3\times k$

7 Calculate the value of each expression when $p = 1, q = 3, r = 0$:

  a $2p+3q+4r$ **b** $p^2+q^2+r^2$ **c** $5pqr$ **d** $4p^2+5q^2$

8 State whether each of the following is true or false:

  a $2a+6 = 18$ when $a = 6$ **b** $5b-5 = 0$ when $b = 1$

  c $x^2+7 = 9$ when $x = 1$ **d** $2y-8 = y$ when $y = 8$

9 Solve these equations for variables on the set of whole numbers:

  a $3x+5 = 17$ **b** $5y-10 = 50$ **c** $8z+1 = 73$

  d $a^2 = 81$ **e** $2a^2 = 18$ **f** $a^2-4 = 21$

  g $7b+13 = 69$ **h** $4c+12 = 12$ **i** $23x = 253$

10 Simplify each of the following where possible:

  a $5x+7x$ **b** $10y-y$ **c** $8z+8z$

  d $4ab-2ab$ **e** $6c-c$ **f** $8d^2+5d^2$

  g $3k+4k-k$ **h** $10q-q+3q$ **i** $x^2+2x^2-3x^2$

  j $5x+4y$ **k** $3x^2-2x$ **l** $5a+3a-2b$

11 Solve the following equations for variables on the set of fractions:

  a $5x = 7\frac{1}{2}$ **b** $2x+1 = 2$ **c** $x-\frac{1}{2} = 4$

  d $5y+y = 9$ **e** $3z-z = 1$ **f** $10k+6k-8k = 20$

12 Find the simplest values or expressions for the volumes of these cuboids:

Length (cm)	4	4	4	c
Breadth (cm)	3	3	b	b
Height (cm)	2	a	a	a

13 Find the simplest values or expressions for the total surface areas of the cuboids in question **12**.

14 x and y are variables on the set $\{1, 2, 3, 4\}$. Find the solution sets of the following sentences:

  a $x+3$ is an even number   **b** $2x+3$ is an odd number

  c $y+3 = 6$ **d** $y+3 = 9$ **e** $y^2 = 1$ **f** $x^2 = 16$

15 Copy and complete these tables:

a

+	x	$2x$	$3x$
$5x$	.	.	.
$4x$	.	.	.
$3x$	.	.	$6x$

b

×	x	$2y$	$3z$
$5x$	.	.	.
$4y$	$4xy$	.	.
$3z$	.	.	.

Revision Exercise 3B

1   Find the value of each of the following when a is replaced by 6:

  a $2a+7$ *b* a^2-36 *c* $3a^2$ *d* $30-4a$ *e* $2a^3$

2   The edge of a skeleton cube made of wire is 5 cm long. Calculate:

  a the total length of wire required

  b the total area of all the 'faces'

  c the volume of the cube.

3   Repeat question 2 for a cube of edge x m.

4 y is a variable on the set $\{0, 2, 4, 8\}$. Find all the possible values of:

  a $4y$ *b* y^2 *c* $2y+12$ *d* $100-y^2$

5   Express each of these in its simplest form:

  a £x in pence   *b* y km in m   *c* z days in hours.

6   Find the cost of:

  a 8 kg at £x per kg   *b* y kg at £y per kg

7   Solve these equations, where the variables are on the set of natural numbers:

  a $10a = 120$ *b* $5b = 75$ *c* $16c = 16$ *d* $8d = 12$

  e $\frac{1}{2}x = 20$ *f* $\frac{1}{4}x = 1$ *g* $\frac{1}{3}x = 8$ *h* $2x^2 = 50$

8   Write each of these in its simplest form:

  a $2\times5\times m\times n$ *b* $3\times k\times7\times n$ *c* $2\times x\times6\times x$

  d $4\times a\times b\times c$ *e* $5\times p\times q\times p$ *f* $2a\times3b\times4c$

9   Find the value of each of the following when $p = 0, q = 1, r = 5$:

  a pqr *b* $6p+4q+2r$ *c* $(p+q+r)^2$ *d* $2q^2+3r^2$

10   Solve the following equations, where the variables are on the set of whole numbers:

  a $3x+1 = 13$ *b* $4y-1 = 19$ *c* $3z-15 = 57$

  d $21-x = 9$ *e* $6x+9 = 51$ *f* $4x+12 = 12$

  g $4a+1 = 8$ *h* $2b-1 = 6$ *i* $4c+2 = 102$

11   Solve the following equations, where x is a variable on the set of fractions:

  a $2x-7 = 8$ *b* $3x-1 = \frac{1}{2}$ *c* $4x+5 = 15$

12   Simplify the following where possible:

  a $8a+6a-10a$ *b* $9b-b$ *c* $5c+6d$

  d $x^2+x^2-x^2$ *e* $3x^2+y^2+2x^2$ *f* $7x^2-x^2$

g $2x+3y+3x+2y$ h $5a+2b-3a-b$ i $8x^2+7y^2-x^2-7y^2$
j $7ab-4ab+9ab$ k $7ab-3cd$ l $6x+1+4x$

13 A cuboid is x cm long, y cm broad, z cm high. Find expressions for:
 a its volume   b the total length of its edges
 c its total surface area.

14 A rectangle is $2x$ m long and x m broad. Its perimeter is 54 m. Find its length in metres, and its area in square metres.

15 Find an expression for the total cost of n books at 35 pence each and n books at 5 pence each. If the total cost was £8, find n.

16 A strip of carpet $2x$ cm long is sewn on to another x cm long, making a strip 2 metres long. What length was each strip of carpet?

17 a and b are variables on the set $\{1, 2, 3, 4\}$, and $a \odot b$ denotes the remainder obtained when a and b are multiplied together and the product is divided by 5. For example, $3 \odot 4$ denotes the remainder when 3×4 is divided by 5, namely 2.

Copy and complete the table shown. Use your table to find the solution set for:

$\odot$	1	2	3	4
1	.	.	.	.
2	.	4	.	.
3	.	.	.	2
4	.	3	.	.

 a $3 \odot x = 2$ b $y \odot 2 = 4$
Is $\odot$ a commutative operation? Give a reason for your answer.

18 The operation $*$ on the set $\{1, 2, 3\}$ is defined by
$$a * b = (a+1) \times (b+1)$$
 a Calculate $1*1, 1*2, 1*3, 2*1, 2*2, 2*3, 3*1, 3*2$, and $3*3$. Is the operation commutative?
 b Calculate $(1*2)*1$ and $1*(2*1)$. Does this *prove* the operation is associative? Try $(1*2)*3$ and $1*(2*3)$.

19 p and q are variables on the set $\{0, 1, 2, 3, 4\}$, and, $p \ominus q$ denotes the *difference* between p and q, i.e. $p - q$ or $q - p$ according to whether p or q is the larger number. If $p = q$ then $p \ominus q = 0$.
 a Make a table to show the values of $p \ominus q$. Is the operation commutative?
 b Calculate $(3 \ominus 2) \ominus 1$ and $3 \ominus (2 \ominus 1)$. Is the operation associative?

20 The operation $\odot$ is defined as in question *17* and the operation $\ominus$ as in question *19* for the set $\{0, 1, 2, 3, 4\}$.
Calculate $3 \odot (1 \ominus 2)$ and $(3 \odot 1) \ominus (3 \odot 2)$. Is the operation $\odot$ distributive over the operation $\ominus$?

Geometry

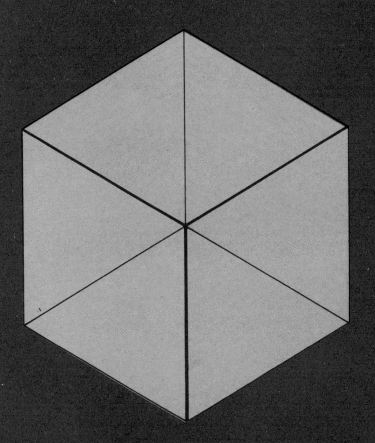

Geometry

Cube and Cuboid

1 Some familiar objects

Exercise 1

1   Figure 1 shows some familiar objects. Can you name them?

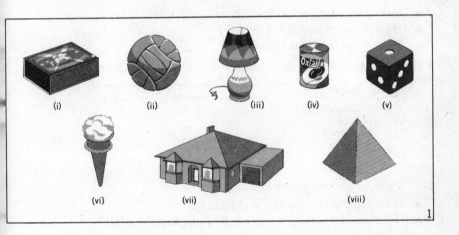

(i)   (ii)   (iii)   (iv)   (v)

(vi)   (vii)   (viii)

1

Special names are given to certain solid shapes in geometry, as shown below:

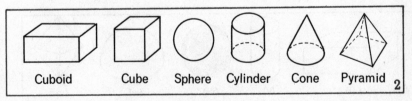

Cuboid   Cube   Sphere   Cylinder   Cone   Pyramid

2

2   Which objects in Figure 1 have these shapes? List your answers under the headings:

Object   Shape

3   Give the names of other objects made up from the shapes shown in Figure 2.

Figure 3 shows the names given to the parts of a *cuboid*.

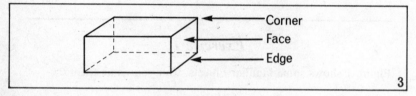

Corner
Face
Edge

3

4 *a*   How many corners has the cuboid?

b   How many edges has the cuboid?

c   How many faces has the cuboid?

5   Copy this table of those shapes shown in Figure 2, and fill it in.

Shape	Number of faces	Number of corners	Number of edges
Cuboid			
Cube			
Pyramid			
Sphere			
Cone			
Cylinder			

6   Think of the classroom as a large box.

a   How many corners, edges and faces has it?

b   What are the various 'faces' of the room usually called?

7   Which of the shapes in Figure 4 have:

a   only straight edges

b   only curved edges?

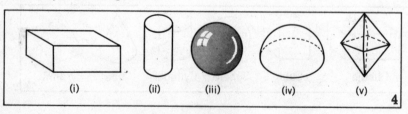

(i)   (ii)   (iii)   (iv)   (v)

4

8   Is it possible to draw a straight line on:

a   a sphere   *b*   a cylinder   *c*   a cone?

9 The world contains objects in many shapes. Name these objects:

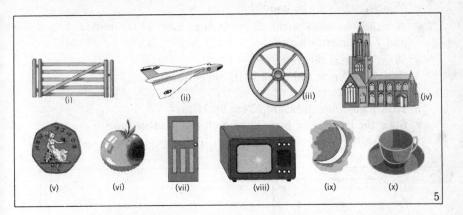

5

10 Make a class collection of some of the following:

a Pictures of interesting shapes used in man-made objects, e.g. buildings, bridges, networks of roads or railways, power stations, atomic energy 'domes'.

b Pictures of interesting shapes in nature, e.g. shells, crystals, leaves, butterflies.

c Pictures of shapes caused by movement, e.g. ripples in a pond, waves on the shore, cloud effects.

2 Walls and bricks

Exercise 2

1 a Which of the following shapes could be used to build a wall without any gaps in it?

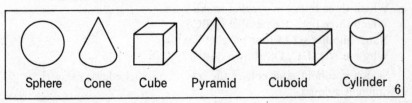

6

b Which of these would be good shapes to use?

c Which would not be good shapes to use?

2   A wall is made of bricks 22 centimetres long, 11 centimetres broad and 7 centimetres high.

a What shape is this brick?

b Sketch the brick and mark in its dimensions.

c How many faces has it?

d How many faces measure 22 cm by 11 cm?

e State the length and breadth of each of the other faces.

Each face of this brick (or cuboid) has the shape of a rectangle.

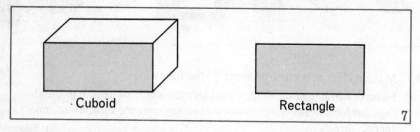

Cuboid   Rectangle

7

3   How many different sizes of rectangular face has the brick described above?

Figure 8 shows a cuboid with all its corners named. Edges which point in the same direction are shown by *parallel lines*.

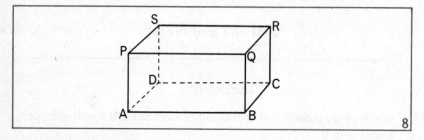

8

4 a Which three lines are parallel to AB?

b Which three lines are parallel to BC?

c Name another set of four parallel lines.

d Why are AD, DC, DS shown dotted?

5   Sketch a cuboid, and colour the sets of parallel edges, using three different colours.

Walls and bricks

If all the edges of a brick (or cuboid) have the same length and all the faces are *squares*, then this special cuboid is called a *cube*.

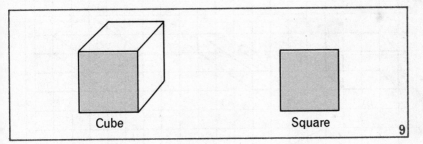

Cube   Square

9

6 a Sketch a cube, and colour sets of parallel edges.

b How many square faces has a cube?

7 Give the names of several objects which have the shape of a cube.

8 In Figure 10 give the names of:

a three lines parallel to BC;

b three lines parallel to AE;

c another set of four parallel lines.

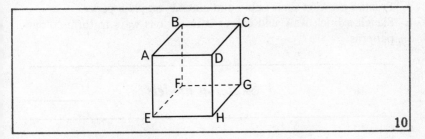

10

9 A brick is 15 cm long and 10 cm broad. Could it be a cube?

10 A brick is 10 cm long and 10 cm broad.

a Could it be a cube?

b Must it be a cube?

c Draw sketches to illustrate your answers.

Figure 11 shows how cubes and cuboids can be drawn on squared paper. Dotted lines indicate edges which are hidden from view.

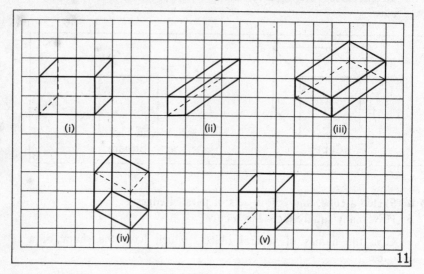

11

11a In which views are you looking down on the shape?
  b In which view are you looking up at the shape?

12a Use your ruler and pencil to draw some cubes and cuboids on squared paper.
  b Practise drawing some cubes and cuboids on plain paper.
  c Sketch a brick wall, and colour it in various ways to form colour patterns.

3 Skeleton models

You can make 'skeleton models' of cubes, cuboids and other shapes using straws joined by lengths of pipecleaner, or dowel rods and Plasticine, or other construction material.

Exercise 3

1 a For the skeleton cuboid in Figure 12(i), how many pieces of rod 13 cm long would you need?

b How many pieces 6 cm long would you need?

c How many pieces 5 cm long would you need?

d Show that altogether you would need almost 1 metre of rod to make the model. Use two methods to calculate your answer, if you can.

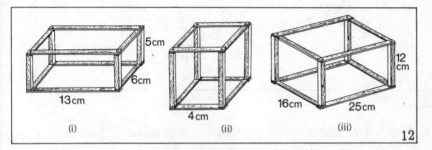

(i)   (ii)   (iii)

12

2 What is the total length of rod required for the skeleton cube shown in Figure 12(ii)?

3 What is the total length of rod required for the skeleton cuboid shown in Figure 12(iii)?

4 What length of rod would be needed

 a for the skeleton cuboid in Figure 13(i)

 b for the pyramid in Figure 13(ii), each of its slanting edges being 12 cm long

 c for the 'house' in Figure 13(iii), formed by putting together the cuboid in (i) and the pyramid in (ii)?

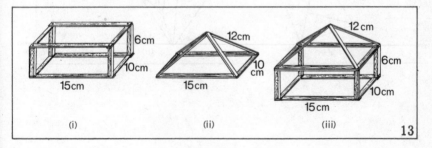

(i)   (ii)   (iii)

13

5 Suppose we made one corner of a cuboid with rods or straws cut and joined as shown in Figure 14(i).

a If we had four such corners, could we fit them together to form a skeleton cuboid?

b Sketch, or make, such a cuboid.

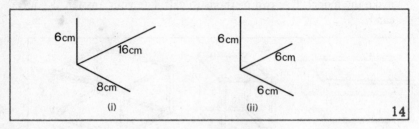

6 *a* Sketch, or make, a cube with one corner as shown in Figure 14(ii).

b What is the total length of its edges?

7 Given twelve pieces of rod could you make a cube or a cuboid if

a all the pieces were the same length;

b eight pieces were of one length and four of another length;

c six pieces were of one length and six of another length?

8 Construct the skeleton models shown in Figure 15, and some others of your own design.

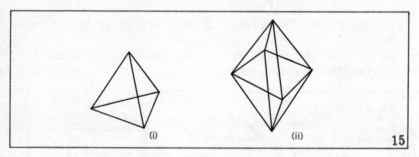

4 Nets of cubes and cuboids

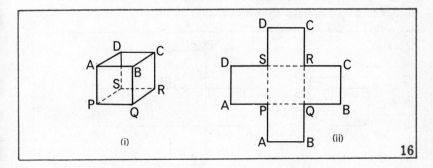

(i)   (ii)

16

Figure 16(i) shows a hollow cube without a lid. If it is cut along the edges AP, BQ, CR and DS it can be folded flat to give the shape shown in Figure 16(ii), which is called a *net* of the cube.

Exercise 4

1 a On 5-mm squared paper draw the shape of Figure 16(ii), making each edge 5 squares (2·5 cm) long.

 b Cut round the outside, and fold it to form a box.

2 Figure 17 shows other ways of placing five equal squares edge to edge.

 a Think of the shapes formed by folding along the dotted lines. Can these nets be folded to form boxes without lids?

 b Copy the nets on squared paper. Cut them out and fold along the dotted lines. Were your answers to *a* correct?

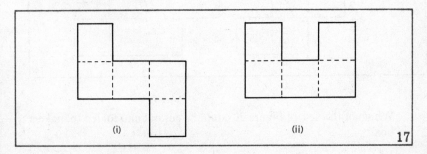

(i)   (ii)

17

3   On squared paper, draw two more nets, one of which will make a box, and one of which will not.

Figure 18(i) shows a box with a lid (a hollow cube). If it is cut along the edges AP, BQ, CR, DS, BA, CD, AD, it can be folded flat to give the shape shown in Figure 18(ii).

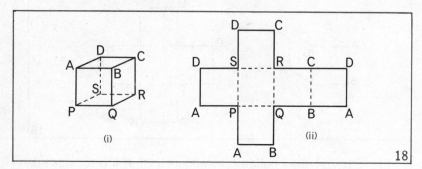

18

4 a On 5-mm squared paper draw the net shown in Figure 18(ii).
  b Cut round the net, and fold it to form a closed box.

5   Figure 19 shows other ways of placing six equal squares edge to edge.

  a Think of the solid shapes formed by folding along the dotted lines. Can these nets be folded to form cubes?

  b Copy the nets on to squared paper. Cut them out and fold them up. Were your answers to a correct?

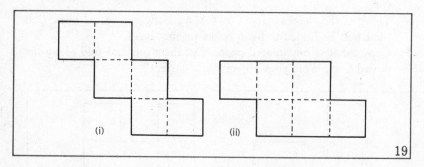

19

6   Which of the nets in Figure 20 could be cut out and folded to make a box?

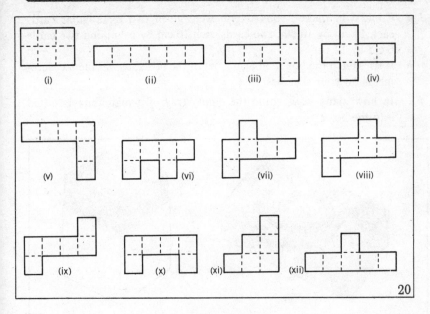

(i)   (ii)   (iii)   (iv)

(v)   (vi)   (vii)   (viii)

(ix)   (x)   (xi)   (xii)

20

Exercise 5

1 a A soap flakes packet is 12 cm long, 5 cm broad and 20 cm high.
Sketch on squared paper, using a simple scale, the shape of each of
the six faces.

  b Two possible nets for the packet are shown in Figure 21(ii) and (iii).
Draw another possible net.

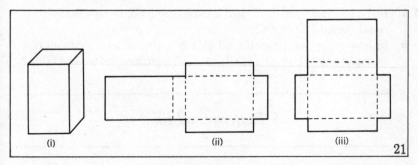

(i)   (ii)   (iii)

21

2 Cut out one of the nets in question *1b* and make the packet. To seal
the closed box, leave a flap on every second edge of the net, and use
a good adhesive (or use adhesive tape).

3 *a*   Could you make a box if you were given two rectangular cards each 14 cm by 10 cm, two cards each 10 cm by 6 cm, and two cards 6 cm by 14 cm?

b   If so, sketch the box, and also a net of the box showing the measurements on your sketches.

4   In how many ways could the empty tray of a matchbox be fitted into the box?

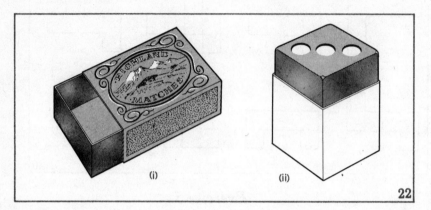

(i)   (ii)

22

5 *a*   If the cubical die just fits the box shown in Figure 22(ii), in how many ways could you fit it in so that the 'three' face shows at the top?

b   In how many ways could you fit it in so that the 'six' face shows?

c   In how many ways altogether could the die be fitted into the box?

6 *a*   If the box in Figure 22(ii) had a close-fitting lid, in how many ways could the lid be put on?

b   In how many ways can the lid of a box shaped like a cuboid (for example, a shoe box, or a chalk box with a rectangular top) be fitted?

5 Patterns and tilings

Floors and walls are often covered with tiles.

Figure 23 shows three patterns formed from squares or rectangles. In each case the tiles are all the same shape and size. They fit

together without leaving gaps, they do not overlap, and the pattern
could be extended as far as you wished.

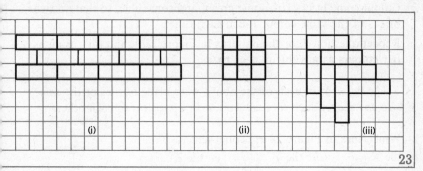

(i)   (ii)   (iii)

23

Exercise 6

1   On 5-mm squared paper draw some tiling patterns of your own
design, using as basic shapes 2×1 or 3×1 rectangles.

2   Sketch some tiling patterns you have seen on buildings, pavements,
etc.; or make a collection of photographs from magazines of tiling
patterns.

3   If a brick is removed from the wall shown in Figure 23(i), in how
many ways can it be replaced?

4 *a*   If the square tiles in Figure 23(ii) are polished on one side only, in
how many ways could a damaged tile be replaced?

  b   If the tiles were polished on both sides, in how many ways could one
be replaced?

5 *a*   In Figure 23(iii) in how many ways could one of the tiles be replaced,
if only one side is polished?

  b   If both sides are polished, in how many ways could one piece be
replaced?

6   Look at Figure 23 again. Which of the three patterns will look the
same if you

  a   turn the page upside down?

  b   turn the page round so that one side goes to the top?

  c   look at the pattern through the page from the back (against the
light)?

7 a Copy the brick wall pattern on to squared paper, and add a few more bricks to the pattern.

   b Colour some bricks red, some green, and some blue, so that no two bricks of the same colour touch each other.

8   Repeat question 7 for the patterns in Figure 23(ii) and (iii), using as few colours as possible in each case.

   The patterns in each of the above were made of tiles of the *same size and shape*. Such tiles are said to be *congruent* to each other.

9   Which tiles in Figure 24 are congruent to

   a tile X   b tile Y   c tile Z?

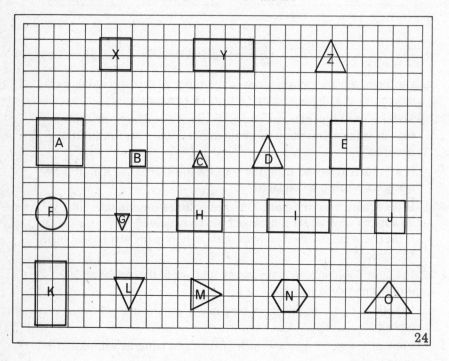

24

On squared paper draw a set of
four congruent squares;
three congruent rectangles.

11a Copy each of the patterns in Figure 25 on to squared paper.
  b Then colour sets of congruent shapes in each pattern, using a
      different colour for each set in a pattern.

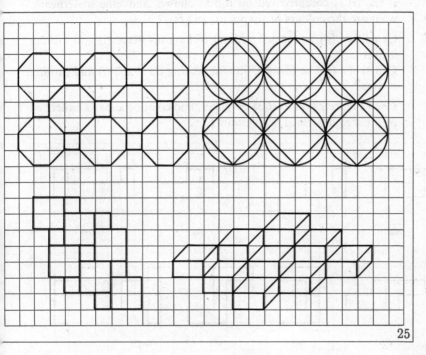

25

---

6 *Half turn symmetry and line symmetry*

Figure 26 shows an L-shaped tile, and two such tiles fitted together
in two different ways to form shapes (ii) and (iii).

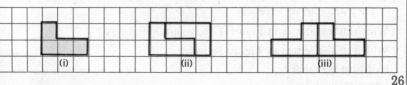

26

Exercise 7

1   If tile (i) is coloured red on top as shown, and blue underneath, what are the colours of shapes (ii) and (iii)?

2 *a* Can shape (ii) be formed without picking up shape (i) and turning it over?

 b Can shape (iii) be formed without picking up shape (i) and turning it over?

3   Copy Figure 27 on to squared paper, and colour the parts either red or blue, as described in question *1*.

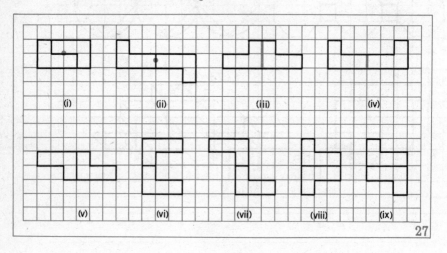

27

The shapes in Figure 27(i) and (ii) fit their outlines when given a half turn about their *centre-points* shown.

They are said to have *half turn symmetry*.

4   Which other shapes in Figure 27 have half turn symmetry?

The shapes in Figure 27(iii) and (iv) fit their outlines when they are picked up and turned over about their *centre lines* shown.

They are said to have *line symmetry*.

5   Which other shapes in Figure 27 have line symmetry?

6   Which of the letters in Figure 28 have

 a half turn symmetry

 b line symmetry?

28

7 *a* Copy the tiles shown in Figure 29.

  b By marking in centre points or centre lines, show which have half turn symmetry and which have line symmetry.

  c Sketch pairs of tiles like (i), placed together in various ways to give shapes with half turn or line symmetry.

  d Repeat *c* for tile (ii).

  e Repeat *c* for tile (iii).

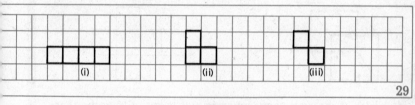

29

8 If you made a collection of interesting shapes in Exercise 1 study these to see whether any of them have the kind of symmetry mentioned above.

Summary

1   Each face of a *cuboid* is a *rectangle*.

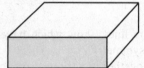

2   Each face of a *cube* is a *square*.

3   Cubes and cuboids each have 3 sets of *parallel* edges. They can be constructed from suitable *nets*.

4   Tiles of the same shape and size are *congruent* to each other.

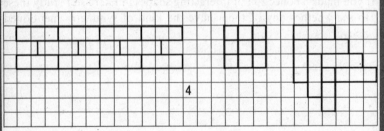

5   Some shapes have *half turn symmetry*.

 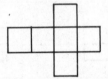

6   Some shapes have *line symmetry*.

Angles

1 Angle as shape

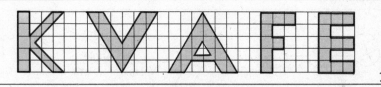

Figure 1 shows part of the board of a child's 'posting' toy. Holes are cut out, and letters made of wood fit into them exactly. To fit the letters in you can turn them over or turn them round.

Exercise 1

1   In how many ways will each letter in Figure 1 fit exactly into its place?

2   In how many ways will each letter in Figure 2 fit exactly into its place?

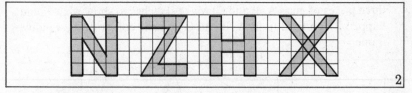

3   Figure 3 shows three versions of the letter O. In how many ways will each fit exactly into its place?

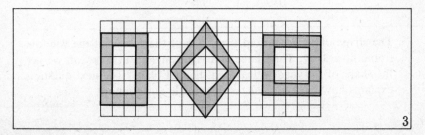

4 *a* Sketch the following letters:

Y P T L O W M I J S

b In how many ways will each fit exactly into its outline?

5 Figure 4(i) shows a 'Meccano' spanner, and Figure 4(ii) shows part of a bicycle spanner.

Sketch the shape of nuts that could be turned easily by these spanners.

(i) (ii) (iii) 4

If shapes are to fit their outlines exactly, the shape of each corner is important. We use *angle* to describe the shape of a corner. To make an angle in our books all we need to draw is a pair of straight lines meeting at a point, as in Figure 4(iii). The straight lines are called the *arms* of the angle, and the point where they meet is the *vertex* (plural *vertices*).

Exercise 2

1 Figure 5 shows the corner of a window frame and the corners of three pieces of glass A, B and C. Make sketches to show what would happen if you tried to fit A, B and C into the corner of the window frame.

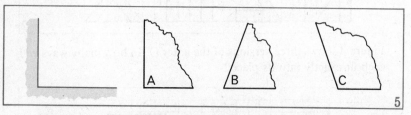

5

The shape of corner A and the shape of the corner of the window frame fit exactly. This shape is a right angle. But how can we get the shape of the window frame in the first place? The next question explains how you can make a right angle quite simply.

2 *a* Take a sheet of paper of any shape. (It does not need to have straight edges.)

  b Make a straight edge by folding the sheet on a flat table or desk (Fold No. 1).

  c Fold the paper again so that one part of the straight edge folds over on to the other part of it (Fold No. 2).

  d Undo Fold No. 2 (see Figure 6(i)).

  e Undo Fold No. 1 (see Figure 6(ii).
     This exercise gives us a good test for a right angle.

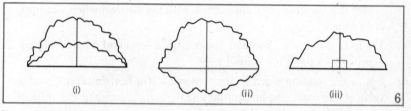

(i)   (ii)   (iii)   6

If two right angles are fitted together their outer arms will be in a straight line; or, two right angles added together make a straight angle (Figure 6(iii)).

Four right angles can be fitted together with no gaps and no overlaps (see Figure 6(ii)).

Notice the symbol for a right angle marked in each right angle in Figure 6(iii).

3   Make a list of objects in the classroom or elsewhere in which right-angle shapes can be seen fitting together.

4 *a* On squared paper sketch part of a brick wall showing right-angle corners fitting together.

  b On your sketch mark some pairs of right-angle shapes which make up straight angles.

5 *a* When the minute hand of a clock points to twelve, at what hours of the day are the hands at right angles?

  b Show the positions of the hands in sketches.

  c Are they at right angles at any other times?

  d At what hour do the hands form a straight angle? Make a sketch of this.

Two lines which form a right angle are *perpendicular* to each other.

6   Draw two perpendicular lines and shade the right-angle shape.

7   If you tied a weight to one end of a piece of string and held the string up by the other end, in what direction would it hang?

        A string suspended from a point, with a weight attached, hangs in the vertical direction through the point.

8   Can you suggest a trade in which vertical lines are used?

9   If you held a pencil at right angles to your string, what name would you give to the direction of the pencil?
        Has this direction any connection with the horizon when you look out to sea?

10   Try to find out what you can about a spirit level and its uses. (There is probably one in the school.)

11a   Fill in the missing word in the sentence. 'If a horizontal line and a vertical line meet, the horizontal line is .... to the vertical line at the point where they meet.'

  b   Name some objects in the room where horizontal and vertical lines meet.

12   How many right-angle shapes are there in the corner of a room?

        Note: A horizontal line is usually represented by a line straight across the page.

Exercise 2B

1 a   Make a sketch of the hands of a clock when they form a straight angle between four o'clock and five o'clock.
  b   Estimate the approximate time when this happens.

2 a   Make a sketch of the hands of a clock when they are at right angles between four o'clock and five o'clock.
  b   Estimate the approximate time when this happens. (There are two possible answers.)

3   You have two equal angle shapes. How could you check whether they are right angles?

4   Say which of the following statements are true and which are false for lines in the classroom:

  a   A horizontal line is perpendicular to a vertical line at the point where they meet.

b Every horizontal line is perpendicular to every vertical line it meets.

c If two lines are perpendicular, one is vertical and the other is horizontal.

d If a line is horizontal, a line perpendicular to it must be vertical.

e It is impossible for two vertical lines to be perpendicular.

2 Angle as amount of turning

Exercise 3

1 Fasten two strips of 'Meccano' at one end as shown in Figure 7. Keep the lower strip steady and turn the other about the joint. What happens to the angle between the strips?

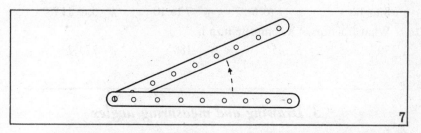

7

2 With the same model, adjust the second strip to make the following angles (or show the angles in sketches):

a a right angle　　　*b* a straight angle　*c* $\frac{2}{3}$ of a right angle

d $1\frac{1}{2}$ right angles

3 Look at a clock or watch face. Through what angle does the minute hand turn when it moves from

a 12 to 3　　　*b* 12 to 6　　*c* 12 to 9　　*d* 12 to 12?

4 Through how many right angles does the minute hand move between 0800 hours and 1030 hours?

5 Name some parts of a motor-car that work by turning.

6 List any objects at home or in the classroom that work by turning.

7 A bicycle wheel turns completely four and a half times. Through how many right angles has it turned?

Until now, we have measured angles in right angles and fractions of a right angle. This is not always convenient, just as it is not convenient to measure all weights in kilogrammes or all lengths in kilometres. To measure smaller angles we need a smaller unit. A complete turn is divided into 360 degrees, written 360°. This unit was chosen a long time ago by the Babylonians.

A complete turn = 360°; a straight angle = 180°; a right angle = 90°.

8   How many degrees are there between the hands of a clock at
  a 3 pm   b 1 pm   c 2 pm   d 4 pm   e 6 pm?
9   Through how many degrees does the minute hand of a clock turn when it moves from
  a 3 to 4   b 2 to 4   c 6 to 10   d 9 to 12
  e 8 to 12   f 1 to 6   g 7 to 10   h 5 to 11?
10 What fraction of a complete turn is
  a 90°   b 45°   c 180°   d 270°?

---

3 Drawing and measuring angles

In order to measure angles in degrees we need a scale for comparison, just as we measure lengths by comparison with the markings on a ruler or tape measure.

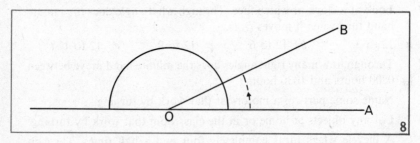

Exercise 4

1 Make a drawing of Figure 8 in your jotter. Take a piece of thread, or hold your ruler on edge, and keep one end fixed at O. Stretch it in the direction OA. Turn the thread or ruler about O into the position OB. What happens to the angle between OA and OB as you turn?

Place the thread or ruler where you think the size of the angle between OA and the thread or ruler would be:

a 45°   b 90°   c 130°   d 180°

Where in your diagram would you mark off a scale in degrees so as to be able to measure the angle between OA and OB?

In Figure 8 we name the angle between OA and OB either AOB or BOA. We always put the letter at the vertex in the middle. The symbol ∠ is used to represent 'angle', and we write ∠ AOB, ∠ BOA, etc.

A protractor is an instrument for measuring angles in degrees. Very often it is semicircular in shape but this is not essential as long as an accurate scale in degrees is marked on it. In fact you may have a protractor on the back of your ruler. If you cannot see how to use your protractor your teacher will guide you. There are two scales on it, and it is important that you use the correct one. So the ability to make a rough estimate of the size of an angle is valuable.

2 a Draw a line OA about 6 cm long and place your protractor approximately in the position of the semicircle in Figure 8.

b Swing the thread or ruler round to make angle AOB = 20°, then 40°, 80°, 90°, 150° and 180°.

3 Use your ruler and protractor to draw the angles below. In each case start by drawing a horizontal arm about 6 cm long.

a ∠ ABC = 45°   b ∠ DEF = 90°   c ∠ GHK = 120°

4 Repeat question 3, starting with a line which is not horizontal.

5 Use your ruler and protractor to draw angles of:

a 20°   b 72°   c 166°

Letter and name the angles yourself.

6 a *Estimate* the size in degrees of each angle in Figure 9, giving your answer like this: ∠ DEF = ...°.

b Use a protractor to measure each angle in Figure 9 in degrees, and
write your answer after your estimate.

c Calculate your error in each case.

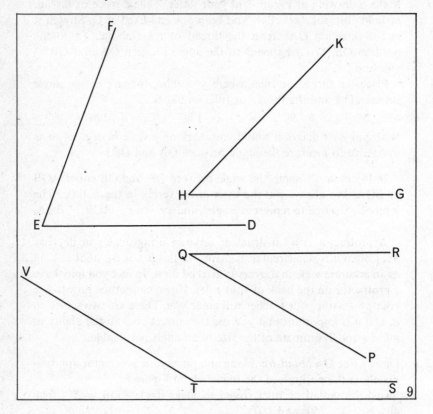

9

7 *a* What is the difficulty about using a semicircular protractor for
measuring all sizes of angles?

   b In question *6*, what rotation would take TS into the position TV in
a *clockwise* direction (i.e. in the direction in which the hands of a
clock turn)?

8   Using your ruler and protractor, draw angles of 5°, 4°, 3°, 2° and 1°.
Do you find it difficult to do this? If so, why?

9 *a* Sketch an angle which you think is 20°, without using your protractor. Measure the angle with your protractor. What was your
error in degrees?

b Repeat for an angle of 60°.

c Repeat for an angle of 130°.

10 Figure 10 shows an electric cooker control with five positions: OFF, 1, 2, 3 and 4 equally spaced round the circle.

a How many degrees are turned through from OFF to 1?

b How many degrees are turned through from 1 to 3?

c How many degrees are turned through from 3 to OFF?

d How many degrees are turned through from OFF right round once to OFF again?

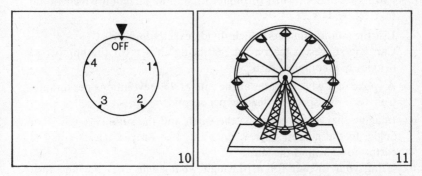

10

11

11 Figure 11 shows the big wheel at a fairground with 12 large supporting spokes, each making the same angle with the next spoke.

a If one car is right at the bottom, through how many degrees does the wheel turn until the next car is at the bottom?

b If one car is right at the bottom, through how many degrees does the wheel turn until the car directly opposite is at the bottom?

c Through how many degrees does the wheel turn if it makes two complete circles?

12 Fill in the blanks in the following:

a An angle of 90° is called a ... angle.

b An angle of 180° is called a ... angle.

c An angle of 360° is called a ... .

d The arms of an angle meet at its ... .

e A quarter turn is the same as ... degrees.

f A half turn is the same as ... degrees.

Exercise 4B

1   Use a protractor to draw an angle of:

   a   210°   *b*   310°

2 a   A spoke of a wheel turns from the position OA to the position OB through an angle of 130° in the anti-clockwise direction. Make a sketch of the two positions.

   b   Calculate the least angle through which the spoke must turn in the same direction to get back to the position OA. Can you think of two more angles turned through in the same direction which would take OB to OA?

3 a   List the whole numbers which divide exactly into 360.

   b   Can you suggest a reason for 360 being chosen for the number of degrees in a complete turn?

4   A spoke of a wheel rotates at the rate of 960 revolutions per minute. Calculate this speed in degrees per second.

5 a   Imagine that a line connects the earth and the sun. Through what angle, to the nearest degree, does this line turn in one day as the earth moves round the sun?

   b   Using your answer for *a*, through what angle does the line turn during June, July and August?

---

4 Acute and obtuse angles

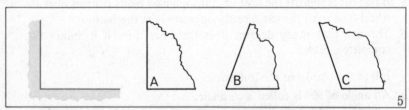

5

In Figure 5 you tried fitting angle shapes into the corner of a window frame. One shape did not fit because it was too 'sharp' and the other because it was too 'blunt'. While we do not describe angles in mathematics as 'sharp' or 'blunt', we do describe them as *acute* or *obtuse*. Look up these words in your dictionary.

We have already seen that it is important to use the correct scale on a protractor. We can help ourselves to do this if we can pick out acute and obtuse angles at sight.

The size of an *acute* angle is between 0° and 90°, and the size of an *obtuse* angle is between 90° and 180°.

Exercise 5

1. State whether each of the following angles is acute or obtuse, without measuring them:

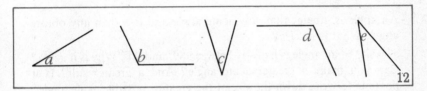

2. List the following angles under the headings acute, obtuse, right, straight:

   160°, 100°, 10°, 90°, 19°, 89°, 91°, 99°, 180°, 179°, 1°, 77°.

3. Do the hands of a clock make an acute, obtuse, right or straight angle at each of the following times?

   a 0900 hours   *b* 1600 hours   *c* 1500 hours   *d* 1100 hours
   e 1800 hours   *f* 0500 hours   *g* 1430 hours   *h* 1715 hours

4. *a* Draw two acute angles and two obtuse angles without using a protractor.
   b Estimate the sizes of the angles in degrees.
   c Measure the angles with a protractor, and write down the error in your estimate in each case.

5. Say whether each of the following is true or false:

   a 89° is an acute angle.
   b 98° is an obtuse angle.
   c The size of an angle depends on the lengths of its arms.
   d There are two right angles in a straight angle.
   e An acute angle can never be equal to an obtuse angle.
   f Two horizontal lines can be perpendicular to each other.

6. In Figure 13 name as many of the following as you can find:

   a acute angles   *b* obtuse angles.

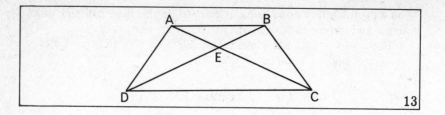

13

Exercise 5B

1   An angle is greater than any acute angle and less than any obtuse angle. What is its size?

2   The size of an angle is halfway between 89° and 90°. Why is it acute? Can you make a greater acute angle? And a greater still? Is it possible to write down the size of the greatest possible acute angle?

3   Are angles of the following sizes acute, obtuse, or neither?

   a $\frac{4}{7}$ of a straight angle   b $\frac{2}{3}$ of a complete turn
   c   1·75 right angles   d $\frac{1}{8}$ of two complete turns
   e $\frac{1}{4}$ of a straight angle   f   179·9°

---

5 *The mariner's compass card. Bearings*

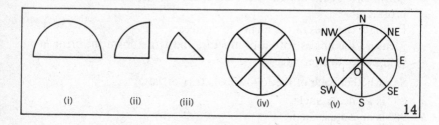

(i)   (ii)   (iii)   (iv)   (v)

14

Exercise 6

1   *First method of constructing a compass card.*

   a   Cut out a circular (or square) piece of paper and fold it over on itself exactly (Figure 14(i)).
   b   Fold it over again (Figure 14(ii)).

c   Fold it over again (Figure 14(iii)).

d   Unfold to obtain the pattern shown in Figure 14(iv).

2   *Second method of constructing a compass card.*

a   Using your compasses, draw a circle with centre O and radius 5 cm.

b   Draw the line WOE across the middle.

c   Draw the line NOS at right angles to WOE.

d   Use your protractor to divide each of the four right angles into two equal parts (Figure 14(v)).

Note: North and South are the main directions, hence NE, NW, SE, SW.

3   What is the size of the least angle in degrees between

a   N and W   *b* N and S   *c* N and SE   *d* SE and SW?

4   Fill in the blanks in the following:

a   The angle between S and … is a right angle. (Give two answers.)

b   The angle between NW and … is a right angle. (Give two answers.)

5 *a* A ship is sailing E and changes course to sail N. Through what angle should it turn?

b   A ship is sailing S and changes to SW. Through what angle does it turn?

c   A ship is sailing NE and changes to S. Through what angle does it turn?

6   You are standing in the classroom facing north.

a   In what direction are you facing after making a half turn?

b   In what directions might you be facing after making a quarter turn?

The above Section describes a simplified form of the mariner's compass. Sailors at one time divided the compass card into thirty-two 'points' by dividing each angle into two equal parts again and

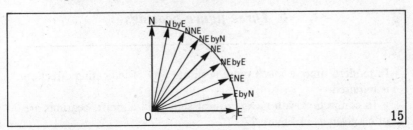

15

yet again. You will find this difficult to do by folding. Figure 15 shows one quarter of the card with the appropriate names.

Sailors were proud of their ability to 'box the compass', i.e. to repeat the points one after the other, 'North, north by east, north north east', and so on.

7 *a* What is the size in degrees of one unit on this scale, e.g. the angle between N and N by E?

 b Why was a more accurate unit of angle unnecessary two hundred years ago?

8   Figure 16 shows five roads meeting at a roundabout where traffic moves clockwise.

 a Which two roads make a straight angle? (Give the road numbers.)
 b Which three pairs of roads meet at right angles?
 c What is the least angle between roads 1 and 2?
 d What is the least angle between roads 2 and 4?

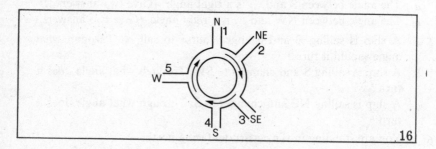

16

9   From an atlas find towns which are approximately

 a NW of London   *b* N of Aberdeen   *c* NW of Edinburgh
 d E of Glasgow   *e* SE of Manchester *f* NE of Perth

6 Three-figure bearings

In modern times a much more accurate way of indicating direction is required.

In connection with navigation of ships and aircraft, bearings are often given as in Figure 17.

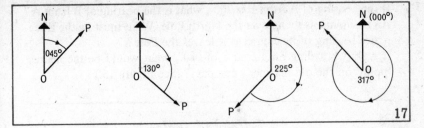

17

An observer at O defines the bearing of an object P by giving the angle between north and the direction OP measured in the clockwise direction.

The bearing is given by three figures, starting from 000°, up to but not including 360°. The direction in which a ship or aircraft is moving may also be given in this way, but the method is not confined to navigation. One hilltop might be described as having a bearing of 176° from another.

Exercise 7

1 What would be the three-figure bearing of the course of an aircraft leaving an airport, and flying

   a east   *b* south   *c* west   *d* north-east *e* south-east?

2 Show in sketches as in Figure 17 the direction of an object from O if its bearing is

   a 060°   *b* 100°   *c* 160°   *d* 280°   *e* 315°

3 A ship is sailing on a course of 215°. Through what angle must it turn to follow a course of 300°? Show the courses in a sketch.

4 An aircraft is heading in a direction 095° and has to change course to 202°. Through what angle must it turn? Illustrate by a sketch.

5 Repeat question 4 when the first course is 338°, and the second is 025°.

Exercise 7B

1 In Figure 16 a driver arrives by road 3 and leaves by road 4. Through what angle does he turn?

2 If the bearing of Edinburgh from Glasgow is 085°, what is the bearing of Glasgow from Edinburgh?

3 If the bearing of A from B is 200°, what is the bearing of B from A?
4 An aeroplane is flying over the North Pole. What must be the three-figure bearing of its course as it leaves the Pole?
5 If a ship is sailing on a course of 080°, what would be the nearest 'point' on the old mariner's compass? (See Figure 15.)

7 Related angles

Supplementary angles

In geometry we often meet angles in pairs. One very simple case is shown in Figure 18(i). If O is a point in AB, it is impossible to draw an angle AOP without at the same time drawing an angle BOP.

We can make a model to illustrate this situation by using two 'Meccano' rods, or by using a wooden rod for AOB and a stretched thread at O for OP, where OP can rotate from OA to OB.

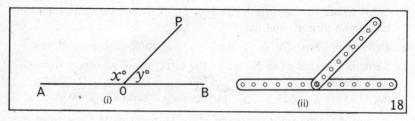

Exercise 8

Questions *1–3* refer to Figure 18(i).

1 What is the sum of the two angles AOP and BOP in degrees?

2 a If $x = 125$, calculate y b If $y = 72$, calculate x
  c If $x = 48$, calculate y d If $y = 99$, calculate x
  e If $x = 0$, calculate y
  f If x increases by 10, what happens to y?

3 a What is the least possible value of y?
  b What is the greatest possible value of y?
  c If $x = y$, what is the value of each?

Two angles which add up to 180° are called *supplementary* angles. Each is the *supplement* of the other.

4 Calculate the supplements of 10°, 100°, 179°, 88°, 90°.

5 In Figure 19, name an angle which is the supplement of

a angle ABE   *b* angle DCE

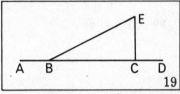
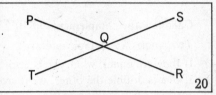

6 a Name four pairs of supplementary angles in Figure 20.

b If $\angle SQR = 65°$, calculate the sizes of the other three angles.

7 Draw two straight lines AB and CD intersecting at O. $\angle AOC = x°$, $\angle BOD = y°$, $\angle BOC = z°$ and $\angle AOD = u°$.

Write down as many equations as you can connecting pairs of x, y, z and u.

8 Two angles AOB and AOC of 43° and 137° respectively are placed so as to lie on either side of a common arm OA. What can you say about

a $\angle BOC$ *b* OB and OC?

Complementary angles
Exercise 9

1 Figure 21 represents a piece of cardboard with a right-angled corner at O. OP represents a piece of thread fastened at O by a drawing pin. OP can be rotated into any position from OA to OB. What is $x+y$?

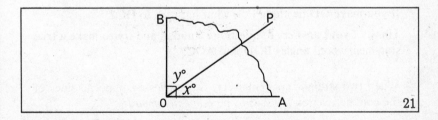

2 *a* If *x* is 25, calculate *y*.
  b If *y* is 62, calculate *x*.
  c When *y* has its least value, what can you say about *x*?

Two angles which add up to 90° are called *complementary* angles. Each is the *complement* of the other.

3 Calculate the complements of 10°, 25°, 89°, 50° and 66°.

4 Two angles are complementary.
  a If they are equal, what size is each?
  b If one is double the other, what size is each?

5 Look up the words *complement* and *supplement* in your dictionary. Try to suggest why these words are used in geometry.

Vertically opposite angles
Exercise 10

1 Figure 22 shows two 'Meccano' rods, AB and CD, bolted at O. Does this model remind you of any cutting instruments you are familiar with?

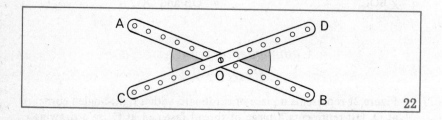

22

2 *a* If you increase ∠BOD by 10°, what happens to ∠AOC?
  b If you increase ∠BOD by *x*°, what happens to ∠AOC?

3 If you make OD lie along OB, what happens to OC?

4 Think of your answers to questions 2 and 3, and try to make a true statement about angles BOD and AOC.

When two straight lines intersect, two angles on opposite sides of the common vertex are said to be *vertically opposite*.

Related angles

Note: Here *vertically* comes from *vertex* and has nothing to do with *vertical.*

Vertically opposite angles are equal.

5 Why is it impossible to have only one pair of vertically opposite angles in a diagram?

6 Figure 23 represents a compass card with the eight principal points marked A, B, C, D, E, F, G and H. Have a competition with your neighbour to see who is first to name all the pairs of vertically opposite angles.

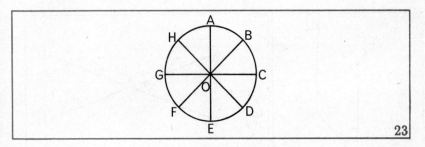

23

Exercise 10B

1 *Calculate* the values of *a, b, c, d, e* and *f* in Figure 24.
State whether each angle is acute or obtuse.

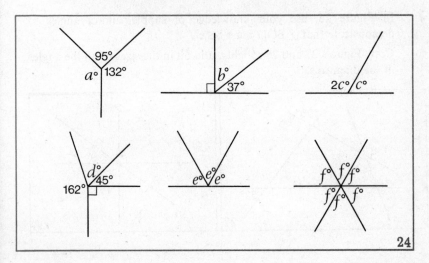

24

2   Is there any difference in meaning between the following two statements?

a   If two angles are complementary, each is an acute angle.

b   If two angles are complementary, neither is an obtuse angle.

    If so, which would you choose as an accurate statement?

3   In Figure 25, C and D are due north of A and B respectively. The bearing of B from A is 050°.

a   Calculate the bearing of A from B.

b   Hence calculate the size of ∠ABD.

c   How are ∠CAB and ∠ABD related?

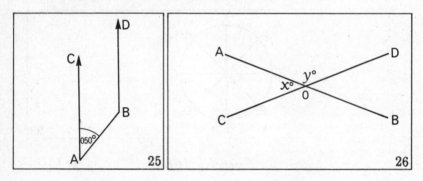

25   26

4   In Figure 26, use your knowledge of supplementary angles to demonstrate that ∠BOD must be $x°$.

5   Copy Figures 27 and 28 roughly, and fill in the sizes of all the angles in the diagrams.

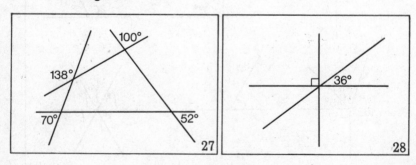

27   28

6   In Figure 29 the four corners are right angles. Copy the diagram and colour or mark in some other way

  a six pairs of vertically opposite angles;
  b four pairs of complementary angles;
  c eight pairs of supplementary angles, not including the right angles.

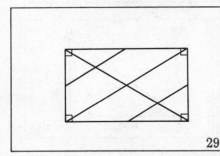

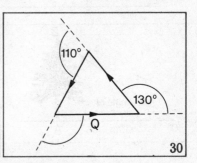

7   In Figure 30 the three lines with arrows represent three straight paths forming a *triangle*. Suppose that, starting at Q you walk round the triangle in the directions of the arrows. The angles that you turn through at two corners are marked.

  a Through what angle have you turned altogether by the time you return to Q?
  b Through what angle do you turn at the third corner?
  c Calculate the sizes of the three angles of the triangle.
  d If the three angles you turn through are $a°$, $b°$ and $c°$, what is the value of $a+b+c$?
  e Write down the sizes of the three angles of this triangle and add them together.

8 Scale drawing. Angle of elevation

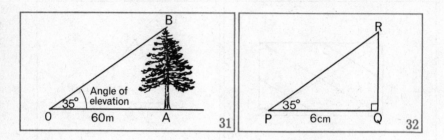

Example 1.—An observer stands at a point O, 60 metres from the foot of a tree. He measures the *angle of elevation* of the top of the tree from O, and finds it to be 35°. We can find the height of the tree approximately by means of a scale drawing.

We cannot measure 60 metres on our page; so we choose a suitable *scale*, and construct a triangle PQR as follows:

a Choose the scale: 1 cm to represent 10 m, so 6 cm represents 60 m.
b Draw a base PQ 6 cm long.
c Make an angle at P of 35°, and an angle at Q of 90°, with arms crossing at R as shown.
d Measure QR. Its length is approximately 4·2 cm, so the height of the tree is about 4·2 × 10 metres, i.e. 42 metres.

It is advisable to make a rough sketch first, but it is not necessary in the rough sketch to attempt to draw trees, houses, ships, etc. In the above question a straight line would do to represent the height of the tree, even in the rough sketch.

The angle of elevation of a point B from a lower point A is the acute angle between the horizontal through A and the direction AB.

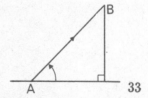

Exercise 11

In the following questions, draw a rough sketch before making the scale drawing in each case.

1   Make scale drawings to find the height of a tree as in Figures 31 and 32 when

   a   OA = 80 m, $\angle$AOB = 30°   b   OA = 150 m, $\angle$AOB = 20°
   c   OA = 20 m, $\angle$AOB = 65°.

2   A ship leaves port X, sails 80 km south to Y, and then 60 km east to Z.

   Make a scale drawing of its course, and find by measurement the distance and bearing of Z from X.

3   A ship leaves port X, sails 60 km west to Y, and then 45 km north west to Z.

   Make a scale drawing, and find by measurement the distance and bearing of Z from X.

4   An aircraft flies 400 km on a bearing of 010° and then 300 km on a bearing of 100°.

   Find the distance and bearing of its starting point from its finishing point.

5   An aircraft flies 1200 km on a bearing of 160°, then 1000 km on a bearing of 270°, and then 800 km on a bearing of 000°.

   Find the distance and the bearing of its starting point from its finishing point.

6   The angle of elevation of the top of a television mast from a point 200 m from its base is 35°.

   a   Make a scale drawing, and find the height of the mast.
   b   From your drawing find also the angle of elevation of the top from a point 120 m from the base.

7   *Project for out of doors*
   Make measurements as in question *1* with tape measure and theodolite or clinometer for a tree, flagpole or building, and find the approximate height by scale drawing and measurement.

   Example 2.—A weather balloon is observed from two places on the same level, 40 km apart. The angles of elevation of the balloon from these places are 42° and 33° respectively. Find the height of the balloon by using a scale drawing.

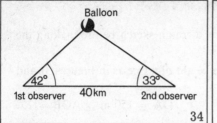

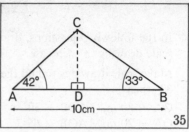

a Choose a suitable scale, e.g. 1 cm to represent 4 km.

b Draw a base AB 10 cm long.

c Make angles of 42° at A and 33° at B with arms crossing at C, as shown.

d Draw CD perpendicular to AB to represent the height of the balloon (using a setsquare, or a protractor).

e Measure CD. Its length is approximately 3·8 cm, so the height of the balloon is about 3·8 × 4 km, i.e. 15·2 km.

Exercise 12

1 Repeat Example 2 for a base line of 80 km, and angles of elevation of 42° and 33°.

2 Repeat Example 2 for a base line of 60 km, and angles of elevation of 38° and 32°.

3 Repeat Example 2 for a base line of 400 km, and angles of elevation of 68° and 75°.

4 Repeat Example 2 for a base line of 15 km and angles of elevation of 55° and 40°.

5 Explain why the method of Example 1 (p. 106) (about the height of a tree) is not suitable in this Exercise.

   Which method would you use to find the height of a church spire? Why?

6 Two boys are situated at points A and B 120 m apart on one bank of a straight stretch of river. They observe a point C on the opposite bank of the river, and measure angles BAC and ABC to be 55° and 65° respectively. Make a scale drawing, and find the width of the river.

7 Repeat question *6* for AB 200 m long, ∠BAC = 45° and ∠ABC = 60°.

Summary

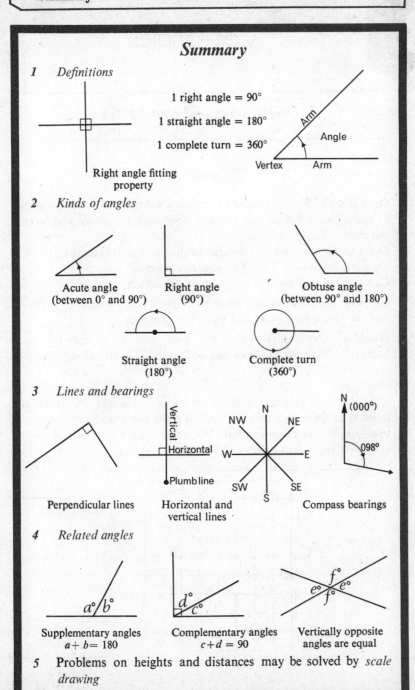

1 *Definitions*

1 right angle = 90°

1 straight angle = 180°

1 complete turn = 360°

Right angle fitting property

Arm

Angle

Vertex   Arm

2 *Kinds of angles*

Acute angle
(between 0° and 90°)

Right angle
(90°)

Obtuse angle
(between 90° and 180°)

Straight angle
(180°)

Complete turn
(360°)

3 *Lines and bearings*

Perpendicular lines

Vertical

Horizontal

Plumb line

Horizontal and
vertical lines

NW   N   NE

W ——————— E

SW   S   SE

N (000°)

098°

Compass bearings

4 *Related angles*

$a°$/$b°$

Supplementary angles
$a + b = 180$

$d°$$c°$

Complementary angles
$c + d = 90$

$e°$ $f°$ $e°$
$f°$

Vertically opposite
angles are equal

5 Problems on heights and distances may be solved by *scale drawing*

Coordinates

1 Origin, axes and coordinates

Exercise 1

1a On a sheet of 5-mm squared paper mark a dot at the corner of one of the squares, but do not let your neighbour see where you have put it.

b Describe the position of your dot so that your neighbour can mark the same point on his sheet of squared paper.

c Can you tell him another way to find the same point?

2 Let your neighbour mark two or three points on his squared paper, and then tell you how to find them on your page.

3 Would you both mark exactly the same point if you were told to 'start from the bottom left corner and count 3 squares along, then 2 squares up'?

So that everyone can find the same points quickly, we take a point O at the corner of a square and draw two lines from O along two sides of the chosen square. We name these lines OX and OY, and mark numbered scales on them as in Figure 1.

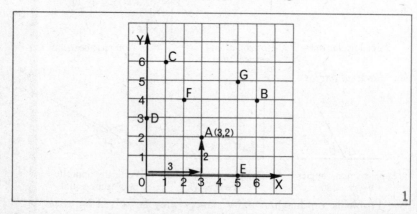

1

O is called the origin, OX the horizontal axis (or the x-axis), and OY the vertical axis (or the y-axis).

We can mark the point A(3, 2) by counting 3 squares from O along OX, and then 2 squares up from OX; *along*, then *up*.

The number pair (3, 2) gives the coordinates of A. The first co-ordinate, 3, is the *x-coordinate* (in the OX-direction) and the second coordinate, 2, is the *y-coordinate* (in the OY-direction).

4 Look at the points named in Figure 1.

a Name the point whose first coordinate is 6.

b Name the point whose second coordinate is 6.

c Write down the coordinates of each of the points B, C, D, E, F, G in the form A(3, 2).

5 On a page of squared paper mark the origin O in the bottom left-hand corner, the axes OX and OY, and the numbered scales on them from 0 to 20.

Plot the following sets of points. Join up the points in each set in the order given, and also join the last point to the first point:

a {A(5, 1), B(8, 6), C(2, 6)}

b {E(14, 2), F(20, 2), G(20, 8), H(14, 8)}

c {P(5, 10), Q(10, 13), R(15, 10), S(12, 15), T(15, 20), U(10, 17), V(5, 20), W(8, 15)}

6 Write down the coordinates of all the corners of the shapes shown in Figure 2, in the form A(2, 2).

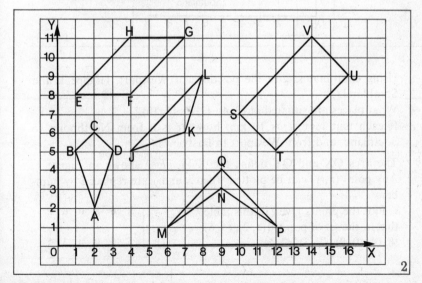

2

7   Join up the separate sets of points below, in the given order, in each
    case also joining the last point to the first, and find the whole object
    that will then be outlined:

 a   {(7, 3), (17, 3), (18, 5), (5, 5)}
 b   {(10, 5), (10, 21)}
 c   {(5, 5), (10, 17), (9, 7)}
 d   {(10, 6), (19, 7), (10, 21)}

8   Figure 3 shows some places marked on an island.

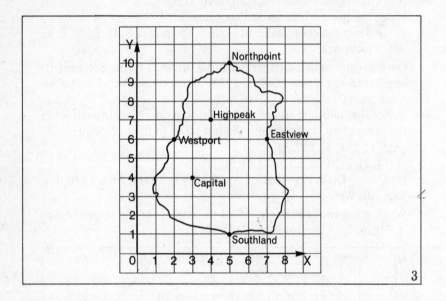

3

 a   Using coordinates, give the positions of Westport, Highpeak and
     Southland.
 b   What places are situated at (3, 4), (7, 6), (5, 10)?
 c   If each unit represents 10 km, how far is it from Northpoint to
     Southland, and from Westport to Eastview?

9   The axes OX and OY define the XOY *plane* shown in Figure 1.
    Any point (x, y) can be plotted when x and y are replaced by num-
    bers.

    Plot the points given in this table,
    that is, the set of points {(1, 2), (6, 6),
    (8, 0), (0, 8), (6, 3), (2, 5)}.

x	1	6	8	0	6	2
y	2	6	0	8	3	5

10   Plot the points given in this table:

x	0	1	2	2½	3	4	5
y	0	2	4	5	6	8	10

11a   Plot the points P(6, 6), Q(6, 10), R(10, 10).
  b   Mark S, the fourth vertex of the square PQRS.
  c   Draw the square and give the coordinates of S.

12a   Plot the points A(15, 5), B(19, 5), C(19, 13).
  b   Mark D, the fourth vertex of the rectangle ABCD.
  c   Draw the rectangle, and give the coordinates of D.

13a   What are the x-coordinates of the points (2, 4), (0, 3), (100, 5)?
  b   What are the y-coordinates of the points (5, 7), (0, 8), (6, 1)?
  c   Give the next four points in the sequence (0, 0), (3, 3), (6, 6), (9, 9)

14a   Draw an interesting shape made up from straight lines on your squared paper, using your ruler and pencil.
  b   Letter the corners, and mark in their coordinates.
  c   Read out the coordinates to your neighbour, and see if he or she can draw the shape you had.

---

2 Sets of points

Exercise 2A

1   Plot the points given by

x	3	6	0	5	2½
y	3	4	5	1	1½

2 a   Plot points to show the set $S = \{(2, 2), (4, 4), (3, 0), (0, 6), (5, 5)\}$.
  b   Is the second member of the given list of elements equidistant from OX and OY?
  c   Which other members of S are equidistant from the axes?

3 a   Write down the coordinates of six points which are equidistant from OX and OY.
  b   If (a, b) is equidistant from OX and OY, what can you say about a and b?

4   Join up the following points in order, and obtain a familiar shape: (3, 2), (13½, 2), (14, 4), (11, 4), (11, 5), (8, 5), (8, 6), (7, 6), (7, 5), (5, 5), (5, 4), (2, 4), (3, 2).

5 *a*   On squared paper, do your best to show the set of all points whose *x*-coordinate is 3.

  b   Is (3, 20) a member of this set? Is (3, 1000)? Is (3, 1 000 000)?

  c   Can you say how many members this set has?

6 *a*   On the same diagram as for question 5, show the set of points whose *y*-coordinate is 4.

  b   Is (4, 4) a member of this set? Is (17, 4)? Is (34 567, 4)?

  c   Can you say how many members this set has?

7   Q is the set of all points with *x*-coordinate 5.
R is the set of all points with *y*-coordinate 8.
Illustrate Q and R, and show what $Q \cap R$ represents.

8   In a 'hunt the treasure' game, the treasure is at the point which belongs to the set of points with *x*-coordinate 3 and also to the set of points with *y*-coordinate 5. Show the position of the treasure on a diagram.

9   A(4, 4) and C(9, 9) are opposite corners of a square ABCD in which AB and DC are parallel to OX and AD and BC are parallel to OY.

  a   Find the coordinates of B and D.

  b   AC is joined and produced its own length to E. Find the coordinates of E.

10   In Figure 4, the lines of the squared paper represent a network of roads. Raiders based at B rob a bank at A. The police set up road

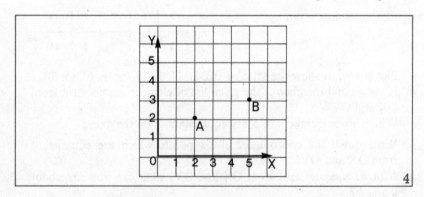

4

blocks at the crossroads (1, 1), (1, 3), (2, 0), (2, 4), (3, 1), (3, 3) and (4, 2).

a Make a diagram to show the above information.

b Can the thieves escape by car to their base?

c What is the least number of right-angled turns they will have to make?

11 Figure 5 shows a side view of a staircase built from blocks whose ends are rectangles 3 units long by 1 unit high.

a Write down the coordinates of A, B, and C.

b Write down the coordinates of the next three points in this sequence.

c If you could continue the sequence, would (30, 10) belong to it? Would (42, 14); or (60, 45); or (600, 200)?

d If $(a, 12)$ is a member of the sequence, what is the value of a?

e If $(24, b)$ is a member of the sequence, what is the value of b?

f If (x, y) belongs to the sequence, which of the following is true?
  (1) $y = 3x$ *(2)* $x = 3y$ *(3)* $3x = 4y$.

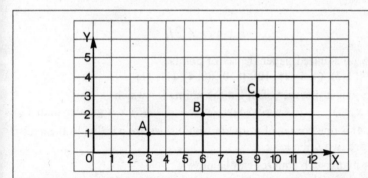

5

12 The game of 'battleships'. You and your opponent each need two 6 by 6 square grids as in Figure 6. On one grid you mark your fleet, on horizontal lines, a battleship joining 4 points, a cruiser 3 points, a destroyer 2 points, and a submarine being represented by 1 point, as in the first grid. You mark your opponent's shots on this grid. On the other grid you record your shots at your opponent's fleet, which he has drawn in a similar way.

Each person gets 5 shots per round. Suppose our first 5 shots are (1, 1), (3, 1), (5, 3), (2, 4) and (4, 5), as shown on the second grid, and that your opponent then tells you that you have scored 2 hits on his cruiser. You would then 'fire' (2, 1) in the next round to finish it off.

The person who sinks the other's fleet first is the winner.

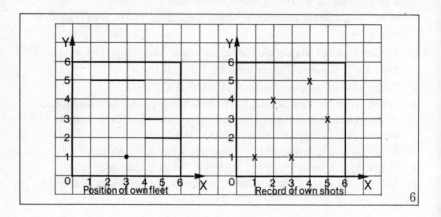

Position of own fleet   Record of own shots

6

Exercise 2B

1 a Show on squared paper the set of points
S = {(2, 3), (4, 4), (5, 10), (8, 0), (8, 4), (11, 11)}.

b If (x, y) is a member of S, list the points for which
(*1*) $x = y$ (*2*) $x = 2y$ (*3*) x is greater than y.

2 A(6, 4) is one corner of a square whose sides are parallel to the axes and are two units long.

a How many such squares can be drawn?

b Write down the coordinates of all the corners of these squares.

3 P is the point (6, 6) and R is (9, 12). PQRS is a rectangle with its sides parallel to the x- and y-axes.

a Find the coordinates of Q and S.

b PR is produced its own length to U. Find the coordinates of U.

c RP is produced its own length to V. Find the coordinates of V.

d What is the area of rectangle UTVW whose sides are parallel to the axes?

4 a Give the coordinates of the next four points in the sequence (2, 2), (4, 4), (6, 6), (8, 8), ...

b Plot the eight points and draw the line through them.

c If (r, s) belongs to the sequence, what can you say about r or s or both?

d Which of the following points lie above the line drawn in *b*?
K(7, 8),   L(10, 10),   M(3, 2),   N(106, 108).

e T(p, q) lies above the line. Which of the following is true?
(*1*) $p = q$ (*2*) p is greater than q (*3*) p is less than q.

5   Q is the set $\{(3, 1), (3, 4), (3, 10), (3, 2)\}$, and T is the set of all points with first coordinate 3.
      Which, if either, is correct: $Q \subset T$ or $T \subset Q$?

6 *a* Draw on squared paper:
(*1*) the set P of points for which $x = 5$
(*2*) the set Q of points for which $x = 10$
(*3*) the set R of points for which $y = 3$
(*4*) the set S of points for which $y = 8$.

b Write down the coordinates of the single point in each of the sets $P \cap R$, $R \cap Q$, $Q \cap S$ and $S \cap P$.

c What are the sets $P \cap Q$ and $R \cap S$?

7 *a* Show the set of points with y-coordinate 10.

b If P(a, b) lies below this line, what can you say about a or b or both?

c If Q(c, d) is a point such that d is greater than 10, where does Q lie relative to the line?

8   A is the set of all points with x-coordinate less than 2. B is the set of all points with y-coordinate less than 4. A player throws a coin, and if it lands in the region $A \cap B$, he wins.

a Shade the winning region in a diagram.

b Is (3, 1) in the winning region? Is (1, 3)? Is (2, 3)?

9 *a* Shade the winning region for the game in question *8* if it consists of that part of the XOY plane containing points whose x-coordinates are greater than 4 but less than 8, and whose y-coordinates are greater than 2 but less than 10.

b Describe the winning region in words.

10*a* O is the origin, A is (2, 0), C (0, 3). Give the coordinates of B, the remaining corner of rectangle OABC.

b Repeat the above with $C_1(0, 7)$ instead of C, then $C_2(0, 10)$ instead of C.

c Is there any limit to the number of rectangles on OA as base?

d If (x, y) is the corner B of such a rectangle, can you say anything about x or y or both?

e If (x, y) is the corner B, and OA is less than the height of the rectangle, can you say anything about x or y or both?

11 Repeat question *11* of Exercise 2A with A the point (4, 3) so that
each rectangle measures 4 units by 3 units.

12 Here is another way to fix the position of a point.

a Draw a line OA 6 cm long across your page, with O near the centre
of the page.

b With centre O, and radii 1, 2, 3, 4, 5, 6 cm, draw circles as shown in
Figure 7. Label the circles 1, 2, 3, 4, 5, 6.

c Using your protractor, draw lines through O at 10° intervals, and
label them 0, 10, 20, 30, . . ., 350.

d Mark the point P[5, 40] at the point of intersection of circle 5 and
line 40. We use square brackets since these are not the usual coordinates.

e Plot the points Q[5, 100], R[5, 160], S[5, 220], T[5, 280], U[5, 340].

f Plot the points B[2, 20], C[4, 20], D[6, 20], E[2, 200], F[4, 200],
G[6, 200].

g Plot other points to form a shape or pattern, and list their 'coordinates'.

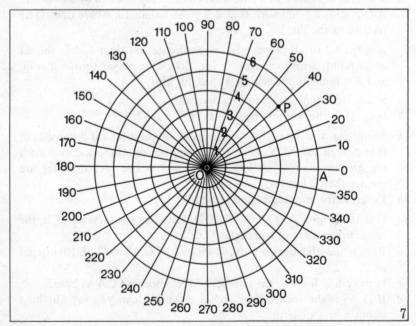

7

13   *Projects*
     Map references are often given by means of coordinate systems.
     Find out what system is used:

a   in your Geography atlas
b   in the A.A. Handbook
c   on an Ordnance Survey map.

Summary

The position of a point is given in the XOY plane by its coordinates. For the point A(3, 2), 3 is the first coordinate or *x*-coordinate, 2 is the second coordinate or *y*-coordinate.

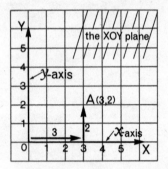

The axes OX and OY are often called Cartesian axes after the seventeenth-century French mathematician Descartes who introduced the idea of coordinates.

Topic to explore

Euler's formula

In geometry you have studied plane-faced solids of various shapes—cuboids, cubes, pyramids, and the 'house' shape on page 73. Copy and complete this table to show the number of faces (F), corners (C) and edges (E) that each has:

Shape	F	C	E
Cuboid			
Cube			
Pyramid			
'House'			

Do you see any connection between the numbers F, C, E in each row of the table?

This relation is called Euler's formula, which was discovered by Descartes in 1640 and proved by another mathematician, Euler, in 1752.

What happens to F, C and E if parts are cut away from the solids in the table? For example, if you cut a corner off a cube you obtain 1 more face, and $3-1 = 2$ more corners, and 3 more edges. Explore this, and other, possibilities.

Revision Exercises

Revision Exercises on Chapter 1
Cube and Cuboid

Revision Exercise 1A

1 Sketch a cube, a cone, and a pyramid.

2 Name one everyday object in the shape of each of these:

 a a cuboid *b* a sphere *c* a cylinder.

3 Sketch and name a solid which has:

 a a curved face only *b* plane faces only
 c both plane and curved faces.

4 Sketch and name a solid which has:

 a no corners *b* one corner only *c* five corners.

5 Figure 1 shows a cuboid with EF = 4 cm, FG = 3 cm, GC = 5 cm. Give the names and lengths of all the other edges of the cuboid.

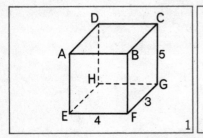

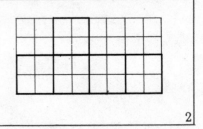

6 Name three sets of parallel lines in Figure 1.

7 A skeleton model of a cuboid is made of thin rods of lengths 4 cm, 5 cm and 6 cm. What is the total length of rod required?

8 36 cm of straw were required to make a skeleton cube. Calculate the length of each edge of the cube.

9 Figure 2 shows the net of an open box. Copy the figure on squared paper, and add one more square to make the net of a closed box.

Cut out and fold the net to check your answer. Can you find other positions for the extra square?

10 Given two rectangular cards 16 cm by 8 cm, and two cards 8 cm by 6 cm, how many more cards would you need to make a box? What size would these cards have to be?

11 State the dimensions of a set of rectangular cards which would make a cuboid with a square base of side 8 cm and height 2 cm.

12 What is the total length of the edges of a cuboid formed from two rectangular cards measuring 3 cm by 5 cm, two measuring 7 cm by 5 cm, and two measuring 7 cm by 3 cm?

13 Figure 3 shows a pattern made from congruent L-shaped tiles. Copy and extend the pattern.

Using four colours, colour the pattern so that two tiles of the same colour do not touch each other.

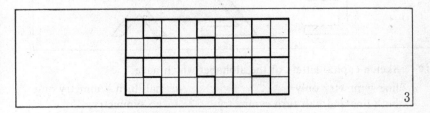

3

14 Repeat question 13 using three colours instead of four.

15 Figure 4 shows a piece of paper folded once, with a coloured shape on the front sheet. The shape is cut out of both sheets, and the paper is then unfolded. Draw a sketch of the hole in the paper, and mark the position of the fold. Check your answer by drawing the coloured shape and cutting it out.

In how many ways can the cut-out shape fit the hole in the paper?

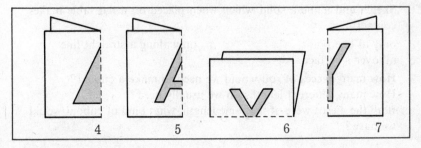

4   5   6   7

16   Repeat question *15* for Figures 5, 6 and 7.

17   Sketch other letters of the alphabet that could be obtained by folding
      once and cutting as in Figures 5, 6 and 7. In each case show the
      position of the fold. In how many ways can each cut-out letter fit
      the hole in the paper?

18   Figure 8 shows three cut-out letter shapes, each one coloured red
      on the front and green on the back.

  a   E is turned over to fit its outline. What colour now shows?
  b   Z is given a half turn to fit its outline. What colour now shows?
  c   In how many ways do E and Z fit their outlines?
  d   In how many ways does X fit its outline with the red surface showing?
  e   In how many ways altogether does X fit its outline?

8

19   Sketch capital letters of the alphabet which have:

  a   line symmetry only   b   half turn symmetry only
  c   both line and half turn symmetry   d   no symmetry.

Revision Exercise 1B

1   Sketch a cuboid, and also a solid made of a cylinder *and* a cone.

2   Sketch and name a solid which has exactly:

  a   one face b   two faces c   three faces d   five faces e   six faces.

3   Sketch and name a solid which has:

  a   no edges   b   only one edge   c   only two edges.

4   Sketch and name a solid which when placed on a flat table makes
      contact:

  a   only at one point   b   only along a straight line
  c   all over a surface.

5 a   How many pieces of rod would we need to make a cuboid?
  b   How many different lengths do we usually need?
  c   If all the pieces were of the same length, what kind of cuboid would
      we have?

d If eight pieces were of one length, and four of another, what could we say about the cuboid?

6 a Is every cube a cuboid?   *b* Is every cuboid a cube?

7 What is the total length of straw needed to make a cuboid with a square base 4 cm long, and with a height of 5 cm?

8 Figure 9 shows part of the net of a closed cubical box. Copy the figure on to squared paper, and add two more squares to complete the net.

Cut out and fold the net to check your answer. Can you find other positions for the two extra squares?

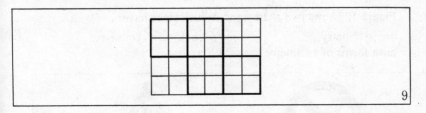

9

9 Given four rectangular cards each 16 cm by 8 cm, what would be the dimensions of two more cards required to make a cuboid? Is there more than one answer? Sketch the cuboid or cuboids.

10 A cuboid has dimensions 4 cm by 5 cm by 10 cm. What would be the dimensions of the cards needed to make a net of the cuboid?

Make two sketches to show how the cards could be arranged on a table to make the two possible nets.

11 Figure 10 shows a pattern of congruent L-shaped tiles. Copy and extend the pattern. Using four colours, colour the pattern so that two tiles of the same colour do not touch each other.

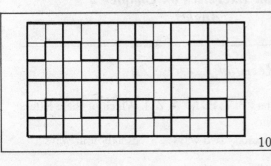

10

11

12   Repeat question *11* using two colours instead of four.

13   Figure 11 shows a sheet of paper folded twice, with a coloured shape on the front surface. The shape is cut out while the paper is folded up as shown. Draw a sketch of the hole in the paper when it is fully unfolded, and mark in the positions of the folds. Check your answer by drawing the coloured shape and cutting it out.

      In how many ways will the cut-out shape fit the hole in the paper?

14   Sketch other letters of the alphabet that could be obtained by folding twice and cutting as in Figure 11. In each case show the positions of the folds. In how many ways can each cut-out letter fit the hole in the paper?

15   Figure 12 shows two road signs. Which signs have:

   a   line symmetry   *b*   half turn symmetry
   c   both forms of symmetry.?

(i)   (ii)

12

16   Study or sketch other road signs, stating which have:

   a   one line of symmetry   *b*   two lines of symmetry
   c   half turn symmetry   *d*   no symmetry.

Revision Exercises on Chapter 2
Angles

Revision Exercise 2A

1   Sketch the capital letters P V H N A Z I. Which of these letters fit into their outlines in:

   a   exactly one way   *b*   exactly two ways   *c*   exactly four ways?

2   Equal angles are fitted together at a point so as to fill up the space round the point without overlapping. What is the size of each angle if the number of angles required is:

   a   4   *b*   10   *c*   15   *d*   24?

3   How many right angles are there altogether on the faces of a cube?

4   A searchlight beam makes an angle of 65° with the horizontal. What angle does it make with the vertical?

5   What is the size of the least angle between the hands of a clock at:

   a   9 pm   *b*   5 am   *c*   3.30 pm?

6   A wheel turns through the following angles. In each case write down the size of the rotation in right angles, and in complete turns.

   a   360°   *b*   720°   *c*   1800°   *d*   1620°

7   What can you say about the size of an angle in degrees which is:

   a   acute   *b*   right   *c*   obtuse   *d*   straight   *e*   a complete turn?

8   Say whether each of the following is true or false:

   a   At 1800 hours the hands of a clock form a straight angle.
   b   All vertical lines in the room are parallel.
   c   All horizontal lines in the room are parallel.
   d   If you are facing south and you make a quarter turn you must be facing west.
   e   An angle of 98° is an obtuse angle.
   f   To double the size of an angle you must double the lengths of its arms.
   g   If a motor-car goes once right round a roundabout it turns through 360°.
   h   A vertical pole standing on horizontal ground casts a horizontal shadow on a sunny day.

9   Calculate *d* in each of the following:

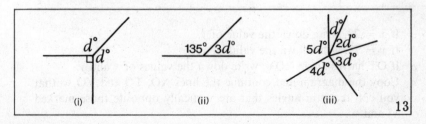

(i)   (ii)   (iii)   13

10 What geographical directions correspond to bearings of:
a 090° **b** 225° **c** 270° **d** 000°?

11 Show accurately on a diagram the directions from a point O given by bearings of 038°, 122° and 300°.

12a What is the size in degrees of the least angle between south-east and south-west?

b A ship sailing south-west changes course through an angle of 45° clockwise. What is its new course?

c If you are facing south-west and you raise your right arm sideways, in what direction does it point?

13 An aircraft flies from A to B on a course of 120°. What course would it have to take to fly from B to A?

14 What is the supplement of:
a 65° **b** 128° **c** 90° **d** $x°$?

15 Say whether the angles in the following pairs are complementary, supplementary, or neither.
a 40°, 140° **b** 60°, 130° **c** 60°, 30° **d** 133°, 47° **e** 72°, 18°

16 $\angle ABC = 32°$. Give the size of an angle which
a is vertically opposite $\angle ABC$
b is complementary to $\angle ABC$
c is supplementary to $\angle ABC$
d makes up a complete turn with $\angle ABC$.

17 In Figure 14, $\angle XOY = 90°$, $\angle XOT = x°$, $\angle TOY = y°$.

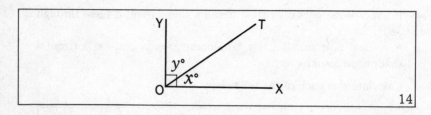

14

a If $x = 20$, write down the value of y.
b If $y = 73$, write down the value of x.
c If OT halves angle XOY, write down the values of x and y.
d Copy the diagram and continue the lines XO, TO and YO so that you can mark in angles that are vertically opposite those marked $x°$ and $y°$.

18 Two straight lines AB and CD cross at O, so that ∠AOC = 72°. What are the sizes of angles COB, BOD and DOA?

19 Two angles are equal. What is the size of each angle if they are:
a complementary   *b* supplementary?

20 The angle of elevation of a church spire from a point 85 metres away horizontally from the base of the spire is 36°. Make a rough sketch to show these facts, and find the height of the spire by means of a scale drawing.

21 The near sides of two buildings are each 50 metres from a point A on the ground. The angles of elevation of the tops of the buildings from A are 35° and 45°. Make scale drawings, and find the difference in height of the two buildings.

22 The angles of elevation of a satellite from two observatories on the earth's surface 55 km apart were 64° and 72°. The satellite is between the observatories and directly above the line joining them. By means of a scale drawing find the height of the satellite at the time.

Revision Exercise 2B

1 In how many ways does the letter X fit into its outline if:
a the smaller angle between the arms is 60°;
b the arms are at right angles to each other?

2 Draw four-sided shapes on squared paper which will fit into their outline in:
a only one way   *b* exactly four ways   *c* exactly eight ways.

3 Equal angles are fitted together at a point so as to fill up the space completely without overlapping. How many angles are required if the size of each is:
a 10°   *b* 15°   *c* 24°   *d* 90°   *e* 120°?

4 How many right angles are there altogether on the faces of a cuboid?

5 What are the sizes of the *two* angles between the hands of a clock at:
a 2100 hours   *b* 2145 hours   *c* 2120 hours?

6 How many right angles, and how many straight angles, are equivalent to:
a two complete turns   *b* 540°   *c* 1620°

7 In Figure 15, four small squares fit together to form a larger square. How many right angles are there in the figure on the next page?

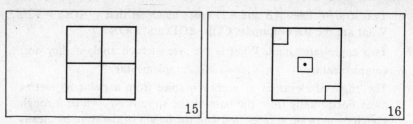

15 16

8 Figure 16 shows three small squares, all the same size, at the centre and corners of a larger square. In how many ways could this figure fit its outline?

Copy the figure and draw in two more small squares so that the number of ways of fitting would be doubled.

9 Say whether each of the following is true or false:

a A vertical line always makes a right angle with a horizontal line it meets in the classroom.

b Two horizontal lines can never be perpendicular to each other.

c If you are facing north-west and you make a quarter turn you must be facing north-east.

d The supplement of an obtuse angle is an acute angle.

e If a bicycle wheel has 48 spokes, the angle between a pair of adjacent spokes is $7\frac{1}{2}°$.

f At 1230 hours the hands of a clock form a straight angle.

10a A cuboid rests with one face on a horizontal table. How many of the faces are horizontal, and how many are vertical?

b Sketch a cube with four of its edges horizontal and none of its edges vertical.

11a What angle does the minute hand of a watch turn through in 1 minute?

b Change a speed of rotation of two turns per minute into a number of degrees per second.

12a If $(90+x)°$ is an obtuse angle, what can you say about x?

b If $(180-x)°$ is an acute angle, what can you say about x?

13a Give the three-figure bearings of north, south and south-west.

b Two aircraft are flying on the same course of 212°. One alters course to 301°, and the other to 179°. Calculate the angle between their courses now.

14a If the bearing of A from B is 345°, what is the bearing of B from A?

b An aircraft is approaching the north pole. What must its course be immediately after it flies over the pole?

15 In Figure 17, AOB is a straight line, $\angle AOP = x°$ and $\angle BOP = y°$.
   a If $x = 3y$, find the values of x and y.
   b If $\angle AOP$ exceeds $\angle BOP$ by 10°, calculate x and y.

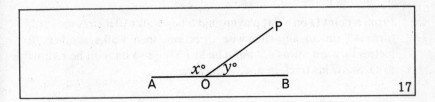

16 Which of the following statements are *always* true for Figure 17?
   a $x = y$ *b* $x + y = 180$ *c* $x - y = 180$
   d As x increases, y increases.
   e As x increases, y decreases by the same amount.
   f When x is doubled, y is halved.

17a What is the supplement of the supplement of $d°$?
   b Two angles are complementary, and each has the same supplement. What is the size of this supplement in degrees?

18 AOB and COD are straight lines making the angles shown in Figure 18. Write down six equations connecting a, b, c, d in pairs.

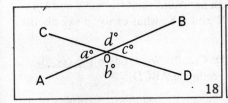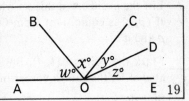

19 In Figure 19 AOE is a straight line.
   a If $\angle AOB = \angle COD = 58°$, and $\angle BOC = 40°$, find the size of $\angle DOE$.
   b If $\angle BOD = 100°$, and $\angle AOB = \angle DOE$, find the size of $\angle BOE$.
   c If $\angle AOC = 70°$, $w = x$ and $y = z$, find the size of $\angle BOD$.
   d If $w = x$ and $y = z$, explain why $\angle BOD$ must be a right angle.

20   From a point 120 metres from the base of a building the angles of
      elevation of two parts of the building are 42° and 58°. By means of
      a scale drawing find how far the second part is above the first.

21   A ship sails 60 km on a bearing 075°, then 50 km on a bearing 180°.
      From a scale drawing find the bearing and distance of the ship from
      its starting point.

22   From a point O on a flat playground a boy walks 10 metres forward,
      turns 45° in an anti-clockwise direction, then walks another 10
      metres forward, turns 45° anti-clockwise, and so on until he returns
      to O. Show his track in a scale drawing.

Revision Exercises on Chapter 3
Coordinates

Revision Exercise 3A

1   Plot the points (1, 5), (4, 3), (11, 2), (19, 2), (18, 5), (18, 7), (20, 7),
      (20, 6), (18, 6), (18, 5), (12, 6), (12, 7), (10, 6), (1, 5). Join the points in
      the order given. Do you recognize the shape?

2 *a* Show on a diagram the set of points
      {(0, 3), (1, 4), (2, 2), (3, 7), (4, 4), (5, 2), (8, 8), (9, 2)}.
  b List the members in this set which are equidistant from OX and OY.
  c If (p, q) is equidistant from OX and OY, what can you say about
      p and q?

3   Mark the points A(1, 4), B(8, 4), C(8, 0). Write down the coordin-
      ates of D, the fourth corner of rectangle ABCD.

4   Plot the points A(2, 3), C(6, 5). A and C are opposite corners of a
      rectangle whose sides are parallel to the axes. Find the coordinates
      of the other two corners.

5 *a* Indicate on a diagram the line given by the set of points whose
      x-coordinate is 7.
  b Is (7, 5) a member of this set? Is (5, 7)?
  c Which of the following points lie to the left of the line?
      (4, 6), (8, 3), (9, 7), (5, 8)

6 a Illustrate the set of points whose second coordinate is 4.

  b Can you say how many members this set has?

  c Does (7, 3) lie above or below the line?

  d If (a, b) lies below the line, what can you say about a or b?

7 P is the set of all points with x-coordinate 7.

  Q is the set of all points with y-coordinate 2.

  Illustrate P and Q, and show what $P \cap Q$ represents.

8 a Plot the five points given by the table:

x	3	4	5	6	7
y	5	6	7	8	9

  b The points (3, 5), (4, 6), (5, 7), (6, 8), (7, 9) lie on the same straight line. Give the coordinates of the next three points on the line whose coordinates are whole numbers.

Revision Exercise 3B

1   Plot the points $(1, 13\frac{1}{2})$, (5, 13), (8, 13), (11, 9), (12, 9), (11, 13), (14, 13), (15, 11), (16, 11), $(15\frac{1}{2}, 13)$, $(17, 13\frac{1}{2})$, $(15\frac{1}{2}, 14)$, (16, 16), (15, 16), (14, 14), (11, 14), (12, 18), (11, 18), (8, 14), (5, 14), $(1, 13\frac{1}{2})$. Join them up in the order in which they are given to obtain a familiar shape.

2 a Show on a diagram the set of points:

  $S = \{(1, 2), (6, 3), (4, 8), (5, 10), (7, 2), (3, 5), (8, 16)\}$.

  b If (x, y) denotes a member of this set, list the points for which $y = 2x$.

  c List the members in this set for which x is greater than y.

3   A sequence of points on a line have x-coordinates 2, 3, 4, 5, 6, 7, ... The first four members of the sequence are the points (2, 6), (3, 9), (4, 12), (5, 15).

  a Write down the next two members of this sequence and then mark all these points on a diagram.

  b If $(a, 105)$ belongs to this sequence, what is a?

  c If $(36, b)$ belongs to this sequence, what is b?

  d If (x, y) belongs to this sequence, what can you say about x and y?

4 a As in question **3**, the first four members of a sequence of points are (3, 8), (4, 11), (5, 14), (6, 17). What are the next two?

  b If $(p, 59)$ belongs to this sequence, what is p?

  c If $(100, q)$ belongs to this sequence, what is q?

d If (x, y) belongs to this sequence, which of the following statements are true?

(*1*) x is greater than y (*2*) y is greater than x
(*3*) y is less than $3x$ (*4*) $3x = y+1$
(*5*) $y = 3x-1$

5 a A is the point (3, 3), B is (7, 3), C is (7, 6), D is (3, 6). What kind of figure is ABCD?

b Another shape BEFC is drawn congruent to ABCD (i.e. the same shape and size as ABCD). Write down the coordinates of E and F.

6 a Indicate on a diagram the line given by the set of all points whose first coordinate is 10.

b If P(x, y) lies to the right of this line, can you say anything about x or y or both?

c If R(a, b) is a point such that a is less than 10, where does R lie relative to this line?

7 a Indicate the line given by the set of all points whose y-coordinate is 5.

b If T(r, s) lies below this line, what can you say about r or s or both?

c If P(l, m) is a point such that m is greater than 5, where does P lie relative to this line?

8 a A is the set of points for which $y = x$. Write down the coordinates of six members of set A, mark them on squared paper, and join them up.

b B is the set of points for which $y = 2x-2$. Write down the coordinates of six members of set B, mark them on the same diagram and join them.

c What on the diagram represents $A \cap B$?

Arithmetic

Arithmetic

The System of Whole Numbers

1 Some sets of numbers

You are already familiar with the set of whole numbers, or counting numbers, used in arithmetic. We call this set W.

The set of whole numbers $W = \{0, 1, 2, 3, ...\}$

In this chapter we will look at some other sets of numbers also, and at the various laws they obey.

A useful set is $\{1, 2, 3, 4, ...\}$, called the set of natural numbers, N.

The set of natural numbers $N = \{1, 2, 3, 4, ...\}$

Which number appears in W, but not in N?

You are also familiar with even and odd numbers.

The set of even numbers $= \{0, 2, 4, 6, ...\}$
The set of odd numbers $= \{1, 3, 5, 7, ...\}$

A prime number is a whole number which is divisible only by itself and 1, but 0 and 1 are not considered to be prime numbers.

The set of prime numbers $= \{2, 3, 5, 7, 11, ...\}$

A square number is formed by multiplying a whole number by itself. For example, $7 \times 7 = 49$, so 49 is a square number. We often write 7×7 as 7^2 ('seven squared').

The set of squares of whole numbers $= \{0^2, 1^2, 2^2, 3^2, ...\}$
$= \{0, 1, 4, 9, ...\}$

In the same way, we write $7 \times 7 \times 7 = 7^3$ ('seven cubed').

The set of cubes of whole numbers $= \{0^3, 1^3, 2^3, 3^3, ...\}$
$= \{0, 1, 8, 27, ...\}$

Sequences.—When we arrange numbers in order according to some rule, the numbers form a *sequence*. Each number is called a *term* of the sequence.

Example 1.   5, 10, 15, 20, 25, ...

We see that each term after the first can be formed by the rule 'add 5'. The next two terms will then be 30 and 35.

Example 2.   100, 97, 94, 91, 88, ...

These can be found using the rule 'subtract 3'. The next two terms will then be 85 and 82.

Example 3.   0, 1, 8, 27, ...

Here we have the cubes of whole numbers, so the next two terms would be 64 and 125.

Exercise 1A

1   List these sets of numbers:

a   The whole numbers less than 10.
b   The odd numbers less than 12.
c   The even numbers less than 12.
d   The natural numbers less than 8.
e   The squares of the whole numbers up to 10.
f   The prime numbers less than 20.
g   The cubes of the whole numbers up to 5.

2   A set can usually be defined in words in several ways; for example, the set {3, 5, 7} could be described as 'the set of odd numbers from 3 to 7' or 'the set of prime numbers greater than 2 and less than 8'.

Write down a description in words of each of the following sets:

a   {0, 1, 2, 3, 4}   *b*   {1, 3, 5, 7, 9}
c   {10, 12, 14, 16, 18}   *d*   {0, 1, 4, 9}

3 *a*   Select any two odd numbers. Add them. Is your answer even or odd?
b   Repeat for two more pairs of odd numbers.
c   Compare your results with the results obtained by the other members of the class.

4 *a*   Write down the first 10 prime numbers.
b   How many of them are even?
c   How many even prime numbers are there?

5   Write down numbers which can fill the spaces to continue these sequences; describe the rule used in each case.

a   1, 3, 5, 7, ..., ...   *b*   20, 19, 18, 17, ..., ...
c   76, 78, 80, 82, ..., ...   *d*   0, 1, 4, 9, ..., ...

6   Write down the meaning of each of the following (e.g. $5^2 = 5 \times 5$), then work out its value:

a $1^2, 9^2, 16^2, 20^2, 10^2, 100^2, 0^2$ *b* $1^3, 2^3, 3^3, 0^3, 10^3, 100^3$

7   Find a suitable missing number for each of these sequences:

 a   0, 2, 4, ..., 8, 10.   b   1, 10, 19, ..., 37.
 c   500, 400, 300, ..., 100.   d   13, 17, 19, 23, 29, ...

8   List the following numbers under the headings *whole, natural,
     even, odd, prime, square, cube.* (Notice that each number can be put
     under several headings.)

                      0, 1, 2, 5, 8, 9, 100

Exercise 1B

1   List the sets defined as follows:

 a   The whole numbers which are greater than 20 but less than 25.
 b   The even numbers which are greater than 90 but less than 100.
 c   The prime numbers between 20 and 30.
 d   The square numbers which are greater than 20 but less than 50.
 e   The first 13 square numbers.
 f   The even numbers which are prime numbers.
 g   The cubic numbers which are less than 200.

2   Write down a description in words of each of the following sets;
     then, as a check, get your neighbour to list the sets using your
     description.

 a   {1, 2, 3, 4}   b   {1, 4, 9}   c   {101, 103, 105, 107, 109}
 d   {100, 121, 144}   e   {7, 11, 13, 17, 19}

3   Write down the meaning (e.g. $6^2 = 6 \times 6$), then the value, of each of
     the following:

 a $0^2, 1^2, 10^2, 100^2, 1000^2$ b $0^3, 1^3, 10^3, 100^3, 1000^3$

4 a Select any two odd numbers. Multiply the larger by the smaller. Is
     your answer even or odd?
 b   Repeat for two more pairs of odd numbers.
 c   Compare your results with the results obtained by the other members
     of the class.

5   Write down numbers which can fill the spaces to continue these
     sequences; describe the rule used in each case.

 a   2, 4, 6, 8, ..., ...   b   4, 9, 16, 25, ..., ...
 c   2, 4, 8, 16, ..., ...   d   1, 2, 3, 2, 3, 4, ..., ...

6   Each of the following lists of numbers contains six terms of a

sequence, and a number which is not a term of the sequence. Give the rule for forming the sequence, and find the 'odd term'.

a   1, 5, 9, 13, 17, 20, 25   b   1, 4, 9, 16, 20, 25, 36

c   91, 84, 77, 70, 63, 56, 48   d   1, 3, 6, 10, 14, 21, 28

7   Find a suitable missing number for each of the following sequences:

a   4, ..., 12, 16, 20, 24   b   4, 9, ..., 25, 36, 49

8   For the sequence 81, 72, 63, 54, ... which of the following numbers could come next? Give the rule involved.

a   43   b   44   c   45   d   46   e   47

9   Add two more numbers to each of the following, stating the rule used to form the sequence:

a   99, 87, 75, 63, ...   b   3, 2, 4, 3, 5, 4, 6, ...

2 Looking at addition

Look at the table, which shows the operation of addition of pairs of numbers belonging to the set {0, 1, 2, 3, 4, 5}.

Second number

Operation table for addition	+	0	1	2	3	4	5
	0	0	1	2	3	4	5 ← Row
	1	1	2	3	4	5	6
First number	2	2	3	4	5	6	7
	3	3	4	5	6	7	8
	4	4	5	6	7	8	9
	5	5	6	7	8	9	10

↑
Column

We can see many interesting facts in the table.

1   First, the sums of all the pairs of numbers form the number square inside the table. How many numbers are there in the square?

2   Can you find a line of 3s in the table? A line of 8s?

3   Can you see a line of odd numbers in the table? Of even numbers?

4   The line through the numbers 0, 2, 4, 6, 8, 10 is called the *main diagonal* of the square. The numbers in the table are distributed *symmetrically* about the main diagonal: a 1 corresponds to a 1, 2 to 2, 3 to 3, and so on.

5   The *rows* of numbers run across the page, and the *columns* run up and down the page. Taking 3 at the left side, and 2 at the top of the table, we find the sum of 3 and 2 where the '3-row' crosses the '2-column'. Check from the table that $3+2 = 2+3$, $4+5 = 5+4$, $1+0 = 0+1$.

6   The order of addition of any two numbers from the set $\{0, 1, 2, 3, 4, 5\}$ does not matter.

In fact, if a and b are any whole numbers,

$$a+b = b+a.$$

This result is called the commutative law of addition.

Some operations are not commutative. Putting on your stockings and then your shoes is not the same as putting on your shoes and then your stockings!

7   Suppose we invent a new symbol * ('star') which means 'double the first number and add the second'.

Then $3 * 2 = 6+2 = 8$. What does $2 * 3$ equal?

Is $3 * 2 = 2 * 3$? Is * a commutative operation like $+$?

The role of zero.—From the table, the following results are also evident:

$$0+0 = 0$$
$$0+1 = 1+0 = 1$$
$$0+2 = 2+0 = 2$$
$$0+5 = 5+0 = 5$$

Extending this to the addition of 0 and any whole number a,

$$0+a = a+0 = a$$

0 is called the identity element for addition.

Exercise 2A

1   From the Addition Table on page 140, write down the numbers:

a   in the first row   *b*   in the first column

c in the third row   d in the sixth column
e in the main diagonal

2 Calculate the sum of the numbers in the first column; in the second column; in the third column.
   Now *write down* the sum of the numbers in each of the other columns.

3 Write down the number where:
   a the second row and the second column cross;
   b the fifth row and the first column cross.

4 Where in this table do you find the sums:
   0+0, and 1+1, and 2+2, and 3+3, and 4+4, and 5+5?

5 From the table, 4+5 = 5+4 illustrates the commutative law of addition. Write down four more results like this from the table that illustrate the law.

6 a Choose any two even numbers, and write down their sum. Is the sum even or odd?
   b Repeat for any two odd numbers.
   c Repeat for a pair of numbers, one of which is even and the other odd.

7 If * ('star') means 'double the first number and add the second', find the values of:
   a 5 * 4 and 4 * 5
   b 10 * 1 and 1 * 10
   c 2 * 0 and 0 * 2
      Can you find values of a and b for which $a * b = b * a$?

8 If □ ('square') means 'square the first number (i.e. multiply it by itself) and add the second', find the values of:
   a 3 □ 2 and 2 □ 3
   b 5 □ 4 and 4 □ 5
   c 10 □ 1 and 1 □ 10
      Can you find values of a and b for which $a □ b = b □ a$?

9 Here is part of a *magic square*, in which each row, column and diagonal has the same sum.

   a What is the sum of the numbers in the first column?
   b Complete the second row, then the main diagonal, the third column, and the first and third rows so that the sum is the same in each case.

10   Copy and complete these magic squares:

2		
7		
6	1	

a

8	1	6
4		

b

8		4
	5	
6		

c

Exercise 2B

1 a If E stands for 'even number' and O for 'odd number', give an example which illustrates E+E = E.

  b Complete the statements   O+O =
                                 E+O =
  and   O+E =
  Give an example to illustrate each.

2 a Find the sum of the numbers in each of the first three rows of the Addition Table printed at the beginning of this section.

  b Hence *write down* the sum of the numbers in each of the remaining three rows.

3   If $\bigcirc$ ('circle') means 'increase the first number by 10 and add the second', find the values of:

  a 12 $\bigcirc$ 5 and 5 $\bigcirc$ 12
  b 58 $\bigcirc$ 0 and 0 $\bigcirc$ 58
  c 11 $\bigcirc$ 9 and 9 $\bigcirc$ 11
  If a and b represent two whole numbers, is $a \bigcirc b = b \bigcirc a$ always true?

4 $\triangle$ ('triangle') means 'cube the first number and add the second'. Choose several whole numbers to replace a and b and say whether $a \triangle b = b \triangle a$ in every case.
       Can you find values of a and b for which $a \triangle b = b \triangle a$?

5   Can you think of other examples of commutative and non-commutative operations?

6   Find the missing number in each of the following squares in order to complete the patterns of rows and columns. For (ii), add each row and column.

2	4	16
3	9	
4	16	256

(i)

16	3	2	13
5	10	11	8
9	6	7	
4	15	14	1

(ii)

In (ii) what is the sum of each row, each column and each diagonal? Such a square of numbers is called a *magic square*.

Now try these for square (ii):

a Find the sum of all the numbers in the first two rows, and then the sum of all the numbers in the last two rows.

b Find the sum of the first and third columns, and then the sum of the second and fourth columns.

Can you discover more things about this magic square?

7   Here is another magic square.

a Add the numbers in each of the two diagonals.

b Add separately (*1*) each row, (*2*) each column.

c Add the numbers in squares A, B, C and D.

d Add the numbers in squares (*1*) E, F, G and H (*2*) I, J, K and L (*3*) E, I, H and L (*4*) J, G, K and F.

A 96	I 11	J 89	B 68
E 88	M 69	N 91	G 16
F 61	O 86	P 18	H 99
C 19	K 98	L 66	D 81

Can you find other four letters in the square for which the sum of the numbers has this same value?

8   Can you make up a 3 × 3 magic square containing the numbers 1, 2, 3, 4, 5, 6, 7, 8, 9?

3 Subtraction as an inverse operation

Suppose that a, b, and c can be replaced by whole numbers.

If $a+5 = 8$, what is the value of a?

If $10+b = 15$, what is the value of b?

If $c+7 = 13$, what is the value of c?

How did you find the value of a? Did you ask yourself, 'What number added to 5 gives 8?' or, 'What is the value of $8-5$?' The answer to each of these is, of course, 3. The first method depends upon addition while the second depends upon subtraction; you can see that by *subtracting* 5 from 8 you can find what number you must *add* to 5 to give 8.

Subtraction is therefore called the *inverse* operation to addition.

The statement $8-3 = 5$ is equivalent to $5+3 = 8$, and we write this as follows: $\qquad 8-3 = 5 \Leftrightarrow 5+3 = 8$

Similarly $\qquad\qquad 15-6 = 9 \Leftrightarrow 9+6 = 15$

Exercise 3A

If the letters represent whole numbers, write down these numbers:

1　$a+7 = 12$　　　　　2　$b+3 = 10$　　　　　3　$c+12 = 20$

4　$15+d = 21$　　　　5　$1+p = 100$　　　　6　$q+10 = 10$

Copy and complete the following, and give an equivalent statement for each one:

7　$12-7 =$　　　　　8　$18-6 =$　　　　　9　$5-4 =$

Subtract the numbers in questions *10*, *11* and *12*; then check your answers by addition.

10　234　　　　　　　11　507　　　　　　　12　827
　　186　　　　　　　　　　119　　　　　　　　　　432
　　———　　　　　　　　　　———　　　　　　　　　———

13　A boy buys records costing 70 pence, and hands over a £1 note. How does the shopkeeper calculate the change?

14a　In the last Section we found that addition was commutative, i.e. $a+b = b+a$. Is subtraction commutative?

b　Is $3-2 = 2-3$? Is $121-97 = 97-121$? Is $a-b = b-a$?

The inverse operation to 'put on your coat' is 'take off your coat'. Give the inverse operation (if it exists) to each of the following:

15 Put on your shoes.

16 Open the window.

17 Dive into the swimming pool.

18 Add 10.

19 Save up 50 pence.

Exercise 3B

If the letters represent whole numbers, write down these numbers:

1	$q+17 = 25$	*2*	$r+19 = 19$	*3*	$p+9 = 16$
4	$25+s = 35$	*5*	$t+79 = 90$	*6*	$88+v = 100$

Copy and complete the following, and give an equivalent statement for each one:

7	$13-9 =$	*8*	$26-8 =$	*9*	$82-12 =$

Subtract the numbers in questions *10* and *11*; then check your answers by addition.

10   1234
    765
  ——

11   5396
  4447
  ——

12 A girl buys a book costing 17 pence, and hands over a 50-pence coin. How exactly does the shopkeeper give her change?

13 An article costing £2·65 is paid for with a £5 note. Write down a possible sequence of change that could be given.

14a By choosing several replacements from W for a and b, find whether the statement $a-b = b-a$ is true or false. Is subtraction commutative?

  b Can you find any replacements for a and b which make $a-b = b-a$ true?

Give the inverse operation (if any) to each of the following:

15 Walk forward five paces.

16 Shut the door and turn the key clockwise.

17   Bake a cake.

18   Fly 20 km north, then 10 km west.

19   Walk down the street and take the first turning on the left.

4 Looking at multiplication

The following table shows the operation of multiplication on pairs
of numbers belonging to the set {0, 1, 2, 3, 4, 5}.

		Second number					
Operation table for multiplication	×	0	1	2	3	4	5
	0	0	0	0	0	0	0
	1	0	1	2	3	4	5
First number	2	0	2	4	6	8	10
	3	0	3	6	9	12	15
	4	0	4	8	12	16	20
	5	0	5	10	15	20	25

As in the case of the addition table, the products are symmetrically
placed about the main diagonal. *There is symmetry about the main
diagonal*, as we have 0, 0; 0, 1, 0; ...; 20, 20, as shown.

From the table, we select the following results:

(a) $2 \times 0 = 0$ $5 \times 0 = 0$ $0 \times 0 = 0$
       $0 \times 2 = 0$ $0 \times 5 = 0$

Any number belonging to the set multiplied by zero, or zero multiplied by any number belonging to the set, gives zero; i.e.

$$0 \times a = a \times 0 = 0,$$

where a represents any member of the set {0, 1, 2, 3, 4, 5}.

(b) $2 \times 3 = 6$ $5 \times 3 = 15$ $4 \times 5 = 20$
       $3 \times 2 = 6$ $3 \times 5 = 15$ $5 \times 4 = 20$

When any two numbers from the set {0, 1, 2, 3, 4, 5} are multiplied together, the order of multiplication does not matter, i.e. *multiplication, like addition, is a commutative operation.*

If we extend the operation of multiplication to the set of whole numbers {0, 1, 2, 3, 4, ...}, the same results hold true.

Thus, for all whole numbers a and b,

(i) $0 \times a = a \times 0 = 0$

(ii) $a \times b = b \times a$

Result (ii) is called the commutative law of multiplication.

The role of 1 *in multiplication.*—From the multiplication table,

$1 \times 1 = 1$

$1 \times 2 = 2 = 2 \times 1$

$1 \times 3 = 3 = 3 \times 1$ and so on.

Extending to multiplication of 1 by any whole number, we have

$1 \times a = a = a \times 1$

1 is called the identity element for multiplication.

Exercise 4A

1 From the Multiplication Table on page 147, write down the numbers:

a in the second row   *b* in the fourth column.

2 *a* List the set of numbers in the main diagonal of the table.

  b Describe this set in words.

3 From the table, $2 \times 3 = 3 \times 2$ illustrates the commutative law of multiplication. Write down four more results like this from the table that illustrate the law.

4 *a* Find the sum of the numbers in each of the first three columns.

  b By examining your results, *write down* the sum of the numbers in each of the remaining three columns.

5 *a* Write down *three* sequences of numbers that you can find in the table.

  b State how each of the three sequences you have chosen is formed.

  c Add two more numbers to each of these sequences.

6 Suppose that □ means 'square the first number and multiply by the second'. Find the values of

a $3 \square 2$ and $2 \square 3$ *b* $5 \square 4$ and $4 \square 5$ *c* $9 \square 0$ and $0 \square 9$

If a and b represent two numbers, is $a \square b = b \square a$ always true?

7   If * means 'double the first number and multiply by the second', find the values of

a   4 * 3 and 3 * 4

b   12 * 5 and 5 * 12

c   100 * 0 and 0 * 100

What do your results suggest about $a * b$ and $b * a$?

Exercise 4B

1 a   Find the sum of the numbers in each of the first three rows of the Multiplication Table set out at the beginning of Section 4 on page 147.

 b   By examining your results, *write down* the sum of the numbers in each of the remaining three rows. Can you give a reason for the results obtained?

2 a   Write down from the table the set P of products which it contains.

 b   In $S = \{0, 1, 2, 3, ..., 25\}$, which numbers belonging to S do not appear in set P? Write down the set of such numbers.

3   If ○ means 'increase the first number by 10 and multiply by the second', find the values of

a   4 ○ 5 and 5 ○ 4

b   25 ○ 1 and 1 ○ 25

c   11 ○ 0 and 0 ○ 11

If a and b represent two numbers, is $a ○ b = b ○ a$ always true?

4   * means 'multiply the first number by three times the second number'. Choose three pairs of numbers as in question *3*, and decide whether or not you think that the commutative law appears to be true for the operation *.

5 *The associative laws of addition and multiplication*

School magazines were being sold. In one day, three pupils sold 87, 66, and 34 magazines respectively. Suppose you were asked to find the total number sold. You would probably add 87 to 66 and then 34 to this sum. This process can be set down as follows:

$$(87 + 66) + 34 = 153 + 34 = 187$$

The brackets tell us that 87 is to be added to 66 first, and then the sum of these to 34.

An easier way, which may have occurred to you, is shown below:

$$87+(66+34) = 87+100 = 187$$

As a second illustration, suppose we have to find the following sum:

$$32+(968+5426)$$
$$32+(968+5426) = 32+6394 = 6426$$

Suppose we insert brackets round the first two numbers instead of the last two.

$$(32+968)+5426 = 1000+5426 = 6426$$

It is evident that the latter way of doing the sum is easier than the former.

The above sums illustrate another law which is true for the addition of whole numbers.

This is the associative law, which can be stated as follows:

$$(a+b)+c = a+(b+c), \text{ for all whole numbers } a, b, c.$$

We usually write $a+b+c$ for these equal sums.

The law can be extended to more than three numbers, and you may have done this when working out long addition sums such as:

(i) 35
   24
   68
   41
   72

(ii) $27+48+23+52$
$= (27+23)+(48+52)$
$= 50+100$
$= 150$

Easy combinations can be found by adding in different orders the numbers in the columns. We feel that what we are doing is correct. That it is correct is a consequence of the commutative and associative laws of addition.

Let us see if there is a corresponding associative law of multiplication which will also give easier ways of obtaining a product.

(i) $(39 \times 5) \times 2 = 195 \times 2 = 390$
$39 \times (5 \times 2) = 39 \times 10 = 390$

(ii) $(798 \times 4) \times 25 = 3192 \times 25 = 79800$
$798 \times (4 \times 25) = 798 \times 100 = 79800$

These products illustrate the associative law for the multiplication of whole numbers, which can be stated as follows:

$(a \times b) \times c = a \times (b \times c)$, for all whole numbers a, b, c.

We write $a \times b \times c$ (or simply abc) for these equal products.

In practice, a combination of both the commutative and associative laws of multiplication often simplifies calculations.

Thus, $50 \times 87 \times 2 = 50 \times 2 \times 87$
$$= (50 \times 2) \times 87$$
$$= 100 \times 87$$
$$= 8700$$

Exercise 5A

Work out the following, using the easiest way you can find:

1 $23 + 93 + 7$ *2* $15 + 38 + 85$ *3* $47 \times 5 \times 2$

4 $4 \times 98 \times 25$ *5* $5 \times (67 \times 20)$ *6* $(897 \times 25) \times 4$

7	*8*	*9*
67	738	849
23	570	608
25	694	784
25	286	377
84	453	532
76		

10 Find the values of:
 a $(12-5)-4$ and $12-(5-4)$
 b $(20-10)-2$ and $20-(10-2)$
 c $(5-3)-2$ and $5-(3-2)$
 Is subtraction associative, i.e. is $(a-b)-c = a-(b-c)$ for all replacements from W for a, b, c?

11 Find the values of:
 a $7 \times 0 \times 83$ *b* $119 \times 52 \times 0$

Exercise 5B

Work out the following, using the easiest way you can find:

1 $82 + 53 + 47$ *2* $26 + 59 + 24$ *3* $(98 \times 25) \times 4$

4	16	5	472	6	845
	93		350		752
	47		906		276
	65		228		463
	85				

7 Choose replacements from W for a, b and c, and check for them whether or not $(a-b)-c = a-(b-c)$ is true.

Do this again for other values of a, b and c.

Is subtraction associative?

Can you find any replacements from W for a, b and c for which $(a-b)-c = a-(b-c)$?

8 Evaluate:

a $97 \times 0 \times 113$ b $999 \times 999 \times 0$

---

6 Division as an inverse operation

Suppose that a, b, and c can be replaced by whole numbers.

If $5 \times a = 20$, what is the value of a?

If $12 \times b = 96$, what is the value of b?

If $c \times 9 = 27$, what is the value of c?

How did you find the value of a? Did you ask yourself 'What number multiplied by 5 gives 20?' or 'What is the value of $20 \div 5$?' The answer to each of these is, of course, 4. The first method depends upon multiplication, and the second upon division; you can see that to *divide* 5 into 20 you can find what number you must *multiply* by 5 to give 20.

Division is therefore called the *inverse* operation to multiplication. The statement $20 \div 5 = 4$ is equivalent to $4 \times 5 = 20$,

$$\text{i.e.} \quad 20 \div 5 = 4 \iff 4 \times 5 = 20$$

Exercise 6A

If the letters represent whole numbers, write down these numbers:

1 $6 \times a = 72$ 2 $9 \times b = 72$ 3 $c \times 11 = 121$

4 $d \times 2 = 0$ *5* $p \times 12 = 132$ *6* $7 \times q = 63$

Copy and complete the following, and give an equivalent statement for each one:

7 $40 \div 8 = \ldots$ *8* $49 \div 7 = \ldots$ *9* $80 \div 2 = \ldots$

Find the answers in *10* and *11*; then check your answers by multiplication.

10 $345 \div 23$ *11* $1369 \div 37$

12   In Section 4 we found that multiplication was commutative, i.e. $a \times b = b \times a$. Is division commutative?

Is $10 \div 2 = 2 \div 10$? Is $15 \div 1 = 1 \div 15$? Is $a \div b = b \div a$?

Exercise 6B

If the letters represent whole numbers, write down these numbers:

1 $8 \times r = 88$ *2* $5 \times s = 60$ *3* $15 \times t = 150$

Copy and complete the following, and give an equivalent statement for each one:

4 $180 \div 12 = \ldots$ *5* $54 \div 9 = \ldots$ *6* $400 \div 20 = \ldots$

7   Find the value of $4674 \div 123$ and check your result by multiplication.

8   Choose three sets of replacements from W for a and b, and check for them whether or not $a \div b = b \div a$.

Is division commutative?

Can you find any replacements for a and b for which $a \div b = b \div a$?

9   Find the values of:
  a $(8 \div 4) \div 2$ and $8 \div (4 \div 2)$
  b $(36 \div 12) \div 3$ and $36 \div (12 \div 3)$
  c $(100 \div 10) \div 5$ and $100 \div (10 \div 5)$
     Is division associative, i.e. is
$$(a \div b) \div c = a \div (b \div c)?$$

Division by zero.—In finding the answer to $20 \div 5$ we had to find the number which when multiplied by 5 gave 20, i.e. $? \times 5 = 20$.

Let us try to find the answer to $20 \div 0$. Here we have to find the number which when multiplied by 0 gives 20, i.e. $? \times 0 = 20$.

There is no such number, and we say that division by zero is *undefined*.

7 Multiples

Consider the part of the multiplication table given below:

×	0	1	2	3	4	5
0	0	0	0	0	0	0
1	0	1	2	3	4	5
2	0	2	4	6	8	10
3	0	3	6	9	12	15
4	0	4	8	12	16	20
5	0	5	10	15	20	25

We notice that corresponding rows and columns contain the same numbers. So we will study the sequences of numbers in some of the rows.

Second row   0, 1, 2, 3, 4, 5.   These are *multiples* of 1.
Third row.   0, 2, 4, 6, 8, 10.   These are *multiples* of 2.
Fourth row.   0, 3, 6, 9, 12, 15.   These are *multiples* of 3.
Fifth row.   0, 4, 8, 12, 16, 20.   These are *multiples* of 4.
Sixth row.   0, 5, 10, 15, 20, 25.   These are *multiples* of 5.

In general, if x is a member of W, where $W = \{0, 1, 2, 3, ...\}$, then multiples of x are products obtained by multiplying x by each member of W.

For example, multiples of 8 are 8×0, 8×1, 8×2, 8×3, 8×4, ...

i.e. 0, 8, 16, 24, 32, ...

Multiples of 2 are given the special name of even numbers.

$$\{0, 2, 4, 6, ...\}.$$

It is clear that every even number is divisible by 2, and this fact provides a test as to whether a given number is even or not.

Whole numbers which do not belong to the set of even numbers are called *odd numbers*. We can therefore consider an odd number to be 1 more or 1 less than some multiple of 2. The set of odd numbers, as we have already seen, is $\{1, 3, 5, 7, ...\}$.

Summarizing, if x is a member of W then $2 \times x$ gives an even number and $(2 \times x) + 1$ gives an odd number. We thus have the *form* taken by even and odd numbers.

Exercise 7

1 Write down the set of multiples of 4, less than 30.

2 Write out the set of multiples of 9, less than 50.

3 Write out the set of multiples of 12, less than 80.

4 List the set of even numbers between 49 and 61.

5 List the set of odd numbers greater than 10 and less than 20.

6 Write down the set of whole numbers between 10 and 20 which are divisible by 2.

7 What is the set of odd numbers divisible by 2?

8 a Write down the first ten multiples of 5.
   b By looking at these numbers, write down a rule for testing whether a number is divisible by 5 or not.

9 a Write down the first ten multiples of 9, spacing them about 1 cm apart.
   b Under each number write down the sum of its figures.
   c Give a rule for testing whether a number is divisible by 9 or not.
   d Is 15246 divisible by 9?

10a Write down any ten multiples of 3.
   b Use the same method as in question 9 to find a rule for telling whether a number is divisible by 3 or not.
   c Is 15246 divisible by 3?

11a Write down the set of multiples of 2 which are less than 20.
   b Write down the set of multiples of 3 which are less than 20.
   c Write down the set of multiples of 6 which are less than 20.
   d Hence write down the set of numbers less than 20 which are multiples of 2 *and* 3 *and* 6.

12 Repeat question *11* for multiples of 5, 10 and 15 which are less than 40.

8 Least common multiple (LCM)

The set of multiples of 6 is {0, 6, 12, 18, 24, 30, 36, 42, 48, ...}.
The set of multiples of 8 is {0, 8, 16, 24, 32, 40, 48, ...}.
The set of common multiples of 6 and 8 is {0, 24, 48, ...}.

The *least common multiple* (greater than 0) of 6 and 8 is 24. We reject 0 as it is common to all such sets.

The least common multiple of two or more numbers is often written as the LCM of the numbers.

Exercise 8

1 a Write down the set of multiples of 2 which are less than 13.
 b Write down the set of multiples of 3 which are less than 13.
 c Hence give the set of common multiples of 2 and 3, less than 13.
 d What is the LCM of 2 and 3?

2 a Give the set of multiples of 3 which are less than 25.
 b Give the set of multiples of 4 which are less than 25.
 c Hence give the set of common multiples of 3 and 4, less than 25.
 d What is the LCM of 3 and 4?

3 a Find the set of common multiples of 5 and 6 which are less than 40.
 b Give the LCM of 5 and 6.

4 a Find the set of common multiples of 4 and 6 which are less than 25.
 b Give the LCM of 4 and 6.

5 a Find the set of common multiples of 8 and 10 which are less than 50.
 b Give the LCM of 8 and 10.

6 Using the answers to questions *1* to *5*, write down the LCMs of:
 a 2 and 3   *b* 3 and 4   *c* 5 and 6   *d* 4 and 6   *e* 8 and 10

Find *by inspection* (i.e. without listing all the multiples) the LCM of each of the following sets of numbers:

7 3, 5	*8* 5, 7	*9* 2, 5	*10* 2, 4
11 3, 6	*12* 4, 5	*13* 6, 8	*14* 8, 12
15 4, 8	*16* 4, 10	*17* 4, 9	*18* 9, 12
19 6, 9	*20* 7, 12	*21* 7, 14	*22* 10, 12

23   2, 3, 4   *24* 3, 4, 6   *25* 6, 12, 36   *26* 5, 10, 15

27 $A = \{$Multiples of 4 which are less than 50$\}$
      $B = \{$Multiples of 6 which are less than 50$\}$
      $C = \{$Multiples of 8 which are less than 50$\}$.
      List all the members of these sets.

      Find   *a* $A \cap B$ *b* $B \cap C$ *c* $A \cap C$ *d* $(A \cap B) \cap C$
      From the members of $(A \cap B) \cap C$ find the LCM of 4, 6 and 8.

28   Repeat question *27* for the multiples of 5, 6 and 15 which are less than 80.

29   One light flashes every 6 seconds, and another flashes every 8 seconds. If they start flashing at the same instant, after how long will they first flash together again? When will they flash together a third time?

30   Repeat question *29* for two lights flashing every 10 seconds and every 12 seconds respectively.

31   Repeat question *29* for three lights flashing every 8, 10 and 15 seconds respectively.

32   One political party holds its annual conference at Brighton every 4 years and another holds its annual conference there every 6 years. If they held their conferences at Brighton in 1970, when will they next be there in the same year?

33   A car has to have new oil every 5000 km, new plugs every 10 000 km and new tyres every 16 000 km. How many kilometres will the car have travelled before all three changes are needed at the same time?

34   Three men A, B and C do nightshift work at a car factory as follows: A does nightshift on every third night, B on every fourth night and C on every fifth night. If they are all on nightshift duty on March 31st, on what date will they next all be on nightshift duty on the same night?

35   A bus stop is served by two bus routes. On one, buses pass every 25 minutes, and on the other every 40 minutes. If one bus on each route leaves the stop at noon, when should buses from the two routes next reach the stop together?

9 Prime numbers and composite numbers

The set of whole numbers W has been divided into two sets:
(i) the set of even numbers, and (ii) the set of odd numbers.

We can divide the set W into subsets in another way. Before doing so, let us consider multiples of numbers once more. Taking multiples of 3, we have the set

$$\{0, 3, 6, 9, 12, 15, 18, 21, 24, ...\}$$

Thus 21 is a multiple of 3 since 21 is divisible by 3. The phrases 'is a multiple of' and 'is divisible by' can therefore be regarded as having the same meaning.

Now if we study the elements of the set

$$W = \{0, 1, 2, 3, 4, 5, 6, 7, 8, ...\},$$

we see that numbers such as 2, 3, 5, 7, 11 are divisible only by themselves and 1. Numbers such as these are called *prime numbers*. The number 1 has only one divisor, namely 1; in this respect 1 is different from 2, 3, 4, . . ., and 1 is regarded as not being a prime number. Also, 0 is not taken to be a prime number. All the other elements of W are called *composite numbers*. It follows that 0, 1, the prime numbers, and composite numbers make up the set of whole numbers. Note that 2 is the only even prime number.

Exercise 9A

The set of prime numbers between 1 and 100 can readily be found using a scheme known as the *Sieve of Eratosthenes*. The method can be extended beyond 100 although it entails rather tedious work.

1 a Write out the numbers from 1 to 100 in order, in 10 rows as shown, using a table 10 cm by 10 cm in size.

Sieve of Eratosthenes

1	2	3	4	5	6	7	8	9	10
11	12	13	14	15	16	17	18	19	20
21	22	23	24	25	26	27	28	29	30
31	32	33	34	35	36	37	38	39	40
41	42	43	44	45	46	47	48	49	50
51	52	53	54	55	56	57	58	59	60
61	62	63	64	65	66	67	68	69	70
71	72	73	74	75	76	77	78	79	80
81	82	83	84	85	86	87	88	89	90
91	92	93	94	95	96	97	98	99	100

b   Cross out 1, which is not a prime number.

c   Cross out all the numbers divisible by 2, except 2 itself, i.e. cross out all the multiples of 2.

d   Cross out all the numbers divisible by 3, except 3 itself.

e   Cross out all the numbers divisible by 5, except 5 itself.

f   Cross out all the numbers divisible by 7, except 7 itself.

The numbers remaining are the prime numbers between 1 and 100.

2 *a*   How many prime numbers are there between 1 and 30?

b   Write out this set of prime numbers.

3   Which of the following are prime numbers?

15, 51, 23, 32, 47, 74, 59, 95, 101

Exercise 9B

1 a   Write down the first six prime numbers.

b   If you add pairs of the prime numbers in *a*, how can you be sure of getting an even number as the sum?

How can you be sure of getting an odd number as the sum?

2   List the following numbers under the headings *prime, composite*:

112, 117, 127, 145, 147, 149, 151.

3   Find the prime numbers between 100 and 110.

4   Why can you not have a set of three consecutive natural numbers (such as 2, 3, 4, or 11, 12, 13) which are all prime numbers?

5   Can the product of two prime numbers be:

a   an even number   *b*   an odd number   *c*   a prime number?

10 Factors

Every whole number can be expressed as the product of two or more numbers, which are called *factors* of the given number.

For example, $7 = 7 \times 1$, so that 7 and 1 are factors of 7. Again, $12 = 12 \times 1$, or 6×2, or 4×3, or $2 \times 2 \times 3$, so that 1, 2, 3, 4, 6 and 12 are all factors of 12.

The factors of 12 which are prime numbers are 2 and 3. These are called the *prime factors* of 12.

Exercise 10A

1 Which of the numbers 2, 3, 4, 5, 8 are factors of:

 a 30   b 45   c 72   d 165?

2 Write down the prime factors of:

 a 10   b 21   c 26   d 35   e 51   f 77

3 Find the largest number which is a factor of each of the following pairs of numbers:

 a 9, 15   b 18, 27   c 12, 24   d 30, 42

4 Which of the numbers 2, 3, 5, 9, 11 are factors of

 a 792   b 3414   c 2222?

Exercise 10B

1 Which of the numbers 2, 3, 4, 5, 8, 9, 10, 11 are factors of:

 a 72   b 60   c 90   d 165?

2 Write down the prime factors of:

 a 94   b 95   c 91   d 30   e 105

3 Find the largest number which is a factor of each of the following pairs of numbers:

 a 24, 30   b 45, 60   c 24, 84   d 66, 110

4 Which of the numbers 2, 3, 5, 9, 11 are factors of:

 a 1980   b 2970   c 4975?

Example 1.—Find the prime factors of (i) 84 (ii) 315.
We first find an obvious factor, and thereafter factorize those factors which are composite numbers.

$$\begin{aligned} \text{(i)} \ 84 &= 12 \times 7 && \text{(ii)} \ 315 = 5 \times 63 \\ &= 4 \times 3 \times 7 && \qquad\quad = 5 \times 9 \times 7 \\ &= 2 \times 2 \times 3 \times 7 && \qquad\quad = 5 \times 3 \times 3 \times 7 \\ &&& \qquad\quad = 3 \times 3 \times 5 \times 7 \end{aligned}$$

We usually list the prime factors in increasing order from left to right.

Example 2.—Express 2808 as a product of prime factors.

$2808 = 4 \times 702$	The working may also	2 | 2808
$= 2 \times 4 \times 351$	be done like this:	2 | 1404
$= 2 \times 4 \times 9 \times 39$		2 | 702
$= 2 \times 4 \times 9 \times 3 \times 13$		3 | 351
$= 2 \times 2 \times 2 \times 3 \times 3 \times 3 \times 13$		3 | 117
$= 2^3 \times 3^3 \times 13$		3 | 39
		13 | 13
		1

Exercise 11A

Express each of the following as a product of prime factors:

1	30	*2*	42	*3*	36	*4*	24
5	60	*6*	54	*7*	64	*8*	63
9	100	*10*	108	*11*	105	*12*	162

Exercise 11B

Express each of the following as a product of prime factors:

1	75	*2*	102	*3*	128	*4*	184
5	209	*6*	180	*7*	242	*8*	243
9	221	*10*	344	*11*	357	*12*	1728
13	1080	*14*	2916	*15*	2106	*16*	1584

11 The distributive law for the whole numbers

Two men A and B work a 40-hour week. A is paid at the rate of 45 pence per hour and B is paid at the rate of 60 pence per hour. Let us work out how much they earned together in one week.

$$\text{Wages earned by both A and B} = (40 \times 45) + (40 \times 60) \text{ pence}$$
$$= 1800 + 2400 \text{ pence}$$
$$= 4200 \text{ pence}$$
$$= £42$$

Can we obtain the answer in a quicker way? It is clear that in 1 hour A and B together earn 105 pence.

Hence, the wages earned by both A and B = 40 × 105 pence
$$= 4200 \text{ pence}$$
$$= £42 \text{ as before}$$

To compare what has been done, let us set out the relation between the numbers involved. We have:

$$(40 \times 45) + (40 \times 60) = 40 \times (45 + 60)$$

Consider another example. We have two boxes of building blocks, each containing one layer. Call them P and Q.

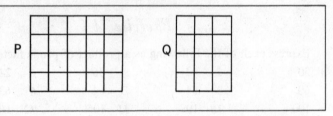

Then, the number of blocks in both P and Q = (4 × 5) + (4 × 3)
$$= 20 + 12$$
$$= 32$$

We can obtain this answer in a quicker way.

The number of blocks in the top row of P and Q is 5 + 3 = 8.

There are 4 rows in each. Hence the number of blocks in both P and Q = 8 × 4 = 32 as before.

As in the first example, we can display the relation between the numbers involved as follows:

$$(4 \times 5) + (4 \times 3) = 4 \times (5 + 3)$$

In both cases, notice that there is a number common to both products, i.e. there is a *common factor* in each.

For the first example, the common factor was 40 and for the second example it was 4. If a calculation shows this form, then it can always be worked out by a quick method. What we have done in this Section is to verify the final law for whole numbers. This law is called the *distributive law* in which multiplication is distributed over addition.

The distributive law states: For all whole numbers *a*, *b*, *c*

$$(a \times b) + (a \times c) = a \times (b + c)$$
or
$$a \times (b + c) = (a \times b) + (a \times c).$$

For simplicity we often write:
$$ab + ac = a(b+c)$$
or $\qquad a(b+c) = ab + ac.$

In practice we very often have to make use of the commutative law first.

Thus $\quad (39 \times 64) + (36 \times 39)$

$\qquad = (39 \times 64) + (39 \times 36) \qquad$ (Commutative law)

$\qquad = 39 \times (64 + 36) \qquad\qquad$ (Distributive law)

$\qquad = 39 \times 100$

$\qquad = 3900$

Exercise 12A

Work out the following, using the distributive law to shorten the calculation. Underline the common factor in each pair of brackets first.

1 $\quad (4 \times 7) + (4 \times 8) = 4 \times (7 + \quad) =$

2 $\quad (3 \times 5) + (9 \times 5) = (3 + \quad) \times 5 =$

3 $\quad (5 \times 11) + (5 \times 9) = 5 \times (\quad + \quad) =$

4 $\quad (6 \times 4) + (19 \times 4) = (\quad + \quad) \times 4 =$

5 $\quad (5 \times 9) + (3 \times 9) = \qquad\qquad$ 6 $\quad (8 \times 5) + (2 \times 5) =$

7 $\quad (9 \times 7) + (9 \times 3) = \qquad\qquad$ 8 $\quad (8 \times 9) + (2 \times 9) =$

9 $\quad (2 \times 13) + (18 \times 13) = \qquad\quad$ 10 $\quad (7 \times 13) + (23 \times 13) =$

11 $\quad (5 \times 31) + (6 \times 31) = \qquad\quad$ 12 $\quad (93 \times 6) + (7 \times 6) =$

Exercise 12B

Work out the following, using the distributive law to shorten the calculation:

1 $\quad (3 \times 19) + (12 \times 19) = \qquad\quad$ 2 $\quad (5 \times 59) + (2 \times 59) =$

3 $\quad (4 \times 7) + (7 \times 8) = \qquad\qquad$ 4 $\quad (3 \times 593) + (2 \times 593) =$

5 $\quad (6 \times 7) + (7 \times 7) = \qquad\qquad$ 6 $\quad (67 \times 41) + (33 \times 41) =$

7 $\quad (45 \times 127) + (273 \times 45) = \qquad$ 8 $\quad (5 \times 98) + (98 \times 95) =$

9 $(987 \times 596) + (13 \times 596) =$

10 $(88 \times 41) + (41 \times 12) =$

11 $\dfrac{(3 \times 9) + (8 \times 9)}{11} =$

12 $\dfrac{22 \times 13}{(13 \times 13) + (9 \times 13)} =$

13 $\dfrac{(5 \times 7) + (4 \times 7)}{(9 \times 3) + (4 \times 9)} =$

14 $\dfrac{(6 \times 5) + (7 \times 5)}{(5 \times 8) + 25}$

Summary of definitions, laws, and principles

The set of *whole numbers* is {0, 1, 2, 3, 4, ...}
The set of *natural numbers* is {1, 2, 3, 4, ...}
The set of *even numbers* is {0, 2, 4, 6, ...}
The set of *odd numbers* is {1, 3, 5, 7, ...}
The set of *prime numbers* is {2, 3, 5, 7, 11, 13, 17, 19, ...}
 Prime numbers are divisible only by themselves and by 1.

1   *The Commutative Laws of addition and multiplication*
$$a+b = b+a \qquad\qquad a \times b = b \times a$$
e.g. $8+24 = 24+8$ e.g. $8 \times 24 = 24 \times 8$

2   *The Associative Laws of addition and multiplication*
$$(a+b)+c = a+(b+c) \qquad (a \times b) \times c = a \times (b \times c)$$
e.g. $(6+4)+3 = 6+(4+3)$ e.g. $(6 \times 4) \times 3 = 6 \times (4 \times 3)$

3   *The Distributive Law*
$$(a \times b)+(a \times c) = a \times (b+c) \quad \text{or} \quad a(b+c) = ab+ac$$
e.g. $(7 \times 3)+(7 \times 5) = 7 \times (3+5)$

4   *The identity element for addition* (0)
$$a+0 = 0+a = a$$
e.g. $5+0 = 0+5 = 5$

5   *The identity element for multiplication* (1)
$$a \times 1 = 1 \times a = a$$
e.g. $8 \times 1 = 1 \times 8 = 8$

6   *Squares and cubes*
a^2 means $a \times a$ a^3 means $a \times a \times a$
e.g. $9^2 = 9 \times 9 = 81$ e.g. $5^3 = 5 \times 5 \times 5 = 125$

7   *The principle for multiplying by* 0
$a \times 0 = 0 = 0 \times a$ If a, b are whole numbers for which
e.g. $7 \times 0 = 0 = 0 \times 7$ $a \times b = 0$, then at least one of a, b is 0.

Later in the course you will see that these laws are similar in form to the laws which govern other number systems you will study. Accordingly, when you have mastered these laws, you will find it easy to carry over your knowledge to other systems of numbers. In fact, you may well be able to discover for yourself, on the basis of these laws, facts which can be applied to algebra and geometry.

Decimal Systems of Money, Length, Area, Volume and Mass

1 Money

British money is based on the *pound*.

One pound (£1) is divided into 100 pence.

One penny (1p) is divided into 2 halfpence.

So we have:

$$235p = 200p + 35p$$
$$= £2 + 35p$$
$$= £2·35$$

Also 45p = £0·45, $98\frac{1}{2}$p = £0·98$\frac{1}{2}$, $3\frac{1}{2}$p = £0·03$\frac{1}{2}$

and £7 = 700p, £3·46 = 346p, £0·87 = 87p, £0·06 = 6p.

Notice that a zero should always be inserted in amounts like the following:

£2·30 (*not* £2·3), and £0·47 (*not* £·47)

Example. $19p \times 100 = 1900p = £19$
$£470 \div 1000 = £0·470 = 47p$

Exercise 1A

1   Express in pounds: 573p, 183p, 72p, 4p, $6\frac{1}{2}$p.

2   Express in pence: £5·87, £2·25, £1·06, £0·09, £0·27$\frac{1}{2}$

3   Add: 53p, $87\frac{1}{2}$p, 42p and $8\frac{1}{2}$p

4   Subtract: 27p from £0·50.

5   I have £1. How much change should I receive after spending 19p and 42p?

6   Multiply the following, giving your answers in £ notation:
   a   13p by 10, by 100, by 1000
   b   £3·35 by 10, by 100, by 1000
   c   29p by 20, by 200
   d   £5·16 by 30, by 300

7   Divide the following, giving your answers in £ notation:
   a   £350 by 10, by 100, by 1000
   b   £425 by 10, by 100

c £5·60 by 10, by 20

d £0·48 by 6, by 8, by 12

8   Find the cost of each of the following:

a 4 spools of film at 28p each.

b 24 bottles of milk at 2½p each.

c 20 magazines at 15p each.

d 4 shirts at £1·45 each.

e 200 pairs of shoes at £2·95 per pair.

9   Find the cost of one article in each of the following:

a 100 articles cost £265.

b 5 articles cost £2.

c 20 articles cost £5·40.

d 1000 articles cost £240.

e 36 articles cost £3·96.

10   How much change is received from a £5 note after buying 6 pairs of socks at 38p per pair and 9 handkerchiefs at 6½p each?

Exercise 1B

1   Bills for £1·36, 55p, 39p and 85p are paid from a £5 note. How much change is received?

2 a Make out a bill for the following:

   3 loaves at 9p each.
   12 eggs at 2½p each.
   2 packets of cornflakes at 10p each.
   4 tins of soup at 7p each.
   3 table jellies at 4p each.

b If a £1 note and a 50p coin are handed over in payment, how much change is due?

c What is the smallest number of coins that could be given for change, and what are they?

3   A school orders 150 books at 40p each, 120 at 60p each, and 50 at 80p each. Find the total cost.

4   The cost of 20 books is £11, but the cost of 100 books is £51·50. How much cheaper is each book when 100 are ordered?

5   A television set costs £75 cash, but can be hired at 40p per week provided an inital payment of £25 is made. If it is hired, how much is paid over the first 2 years? How much is paid over the first 3 years?

6 A tape recorder costs £36 cash. It can be bought with an initial payment of one quarter of the cash price, and 9 monthly payments of £3·25. How much dearer is this way of buying the recorder?

7 Motor-car tyres cost £7·30 each. Find the cost of a set of four tyres.
   A firm advertises that a set of four of their remould tyres will save a motorist £11·36. What is the cost of each remould tyre?

8 A household used 1500 units of off-peak electricity at 0·4p per unit, and 600 units at 0·65p per unit. What was the total bill?

9 The tickets at a school concert cost 12½p and 25p each, and the total takings were £73·25. If 350 of the cheaper tickets were sold, how many of the dearer ones were sold?

10 A motorist travelled 15000 km in a year, and he estimated that his running costs for petrol, oil, tyres and repairs were £324. How much was this cost per kilometre, to the nearest penny?

---

2 Length

Most countries in the world now use a metric system of measures. In this system, the basic units are the *metre* for length, the *kilogramme* for mass, and the *second* for time.

On a metre stick, you will see that 1 metre is divided into 100 centimetres.

On your ruler, you will see that 1 centimetre is divided into 10 millimetres.

For greater lengths the kilometre is used, 1 kilometre being 1000 metres.

Using the abbreviations mm for millimetre, cm for centimetre, m for metre, and km for kilometre, we have:

$$1 \text{ cm} = 10 \text{ mm}$$
$$1 \text{ m} = 100 \text{ cm} = 1000 \text{ mm}$$
$$1 \text{ km} = 1000 \text{ m.}$$

The metre was introduced in France in 1790, shortly after the French Revolution. It was defined to be one ten-millionth part of the distance along a meridian from the North Pole to the Equator, and was given by standard-length metal bars in most countries.

Now the metre is measured in terms of the wavelength of the orange-red line of the krypton gas spectrum (1 metre = 1 650 764 times this wavelength).

Exercise 2

1 Draw lines in your notebook of the following lengths:

a 3 cm   *b* 10 cm   *c* 8 mm   *d* 5 cm 5 mm   *e* 1 cm 1 mm

2 Which metric unit would you use to give the lengths of:

a a book   *b* a room   *c* a desk   *d* a pencil point
e a blackboard   *f* a fingernail   *g* your shoe   *h* your country?

3 Measure the length and breadth of this page, giving your answers in cm and mm.

4 Measure the length and breadth of your desk.

5 Write down your estimate of the following lengths, using the most suitable metric unit:

a the height of the classroom door
b the width of the classroom door
c the length and breadth of the classroom
d the width of the blackboard
e the distance from your school to the nearest town centre
f the distance from your town to the capital of the country.
Now measure the lengths in *a* to *d*, and compare with your estimates. How can you find the lengths in *e* and *f*?

6 a Draw a rectangle 8 cm long and 5 cm broad on squared paper.
  b Calculate its perimeter (the distance right round the edge).

7 a Draw a square of side 6 cm on squared paper.
  b Calculate its perimeter.

8 Copy and complete the following tables:

a Rectangles   *b* Squares

length	breadth	perimeter
5 mm	3 mm	
7 cm	5 cm	
15 m	7 m	
350 m	200 m	
9 cm		32 cm

length of side	perimeter
11 cm	
22 m	
8 m 50 cm	
	500 mm
	6 m 40 cm

9   A terrace in a new town shopping centre is in the shape of a rectangle, 50 m long and 12 m broad. It is to be paved with square slabs of side 50 cm.

a   How many slabs can be placed in a row along the length?
b   How many such rows would be needed?
c   How many slabs are required altogether?
d   What will the total cost be, if one slab costs 12p?

10   Various instruments are used to measure length. Can you give the names of some of them? Think of very small and large lengths, as well as the more common lengths.

11   *Some suggestions for practical work:*

a   Measure the length and breadth of the playground, or a games pitch, in metres.
b   Measure the heights of a group of pupils in your class, and calculate their average height in centimetres (i.e. the sum of the heights of the pupils, divided by the number of pupils).
c   Mark out a length of 50 m in the playground or corridor, using a measuring tape.

Then count the number of steps you take in walking 50 m. How many steps would you take for 1 km? Pace out 1 km on your way home.

*   *   *

1 cm = 10 mm, so 5·3 cm = 5 cm 3 mm = 53 mm;
and 82 mm = 8 cm 2 mm = 8·2 cm.
1 m = 100 cm, so 4·72 m = 472 cm; (compare £4·72 = 472p,
and 23 cm = 0·23 m.   and 23p = £0·23)
1 km = 1000 m, so 2·5 km = 2500 m;
and 5468 m = 5·468 km.

Example 1.   8 m 36 cm + 3 m 4 cm =   836 cm   *or*   8·36 m
                                       + 304 cm   + 3·04 m
                                                   _____   _____
                                       = 1140 cm   = 11·40 m

Example 2.   4 cm 7 mm − 2 cm 9 mm   = 47 mm   *or*   4·7 cm
                                       − 29 mm   − 2·9 cm
                                                   _____   _____
                                       = 18 mm   = 1·8 cm

Exercise 3

1   Express in mm: 4 cm, 47 cm, 4·7 cm, 11·5 cm

2   Express in cm: 4 m, 47 m, 4·7 m, 4·73 m

3   Express in cm: 60 mm, 68 mm, 6 mm, 256 mm

4   Express in m: 700 cm, 750 cm, 750 mm, 2897 mm

5   Express in m: 8 km, 19 km, 19·5 km, 1·27 km

6   Express in km: 5000 m, 15 000 m, 3250 m, 700 m, 75 m, 7 m, 2 km 125 m, 5 km 5 m

7   Add the following, and give each answer in two forms, as in the worked examples:

    a   8 cm 8 mm, 6 cm 6 mm, 5 cm 5 mm
    b   12 cm 3 mm, 1 cm 1 mm, 15 cm 9 mm
    c   5 m 60 cm, 4 m 40 cm, 2 m 15 cm
    d   12 m 25 cm, 8 m 88 cm, 16 m 73 cm

8   Subtract the following, and give each answer in two forms:

    a   3 cm 4 mm from 8 cm 2 mm
    b   5 cm 8 mm from 12 cm 1 mm
    c   1 m 75 cm from 3 m 16 cm
    d   7 m 16 cm from 16 m 8 cm

9   Copy and complete these tables:

    a   *Rectangles*   b   *Squares*

length	breadth	perimeter
25 cm	15 cm	
8 mm	5 mm	
15 cm		42 cm
6 mm		2 cm

length of side	perimeter
5 mm	
37 cm	
	36 mm
	42 cm

10   Some model cars are 56 mm long.

    a   If 12 of them are placed front to back, how far will they stretch?
    b   If a line of the cars stretches for 39 cm 2 mm, how many cars are there?

11   A certain wavelength is 245 mm. Find the length of 20 waves.

12   The average thickness of a concert ticket is $\frac{1}{4}$ mm. 128 tickets are placed in a pile. What is the height of the pile?

13   What length of wire would you need to make the following skeleton models?

  a A cuboid 20 cm by 15 cm by 10 cm.

  b A cube of edge 12·5 cm.

14   A child has a set of 10 cubes of different sizes. The edges of the cubes are 2 cm, 3 cm, 4 cm, and so on, in length,

  a What is the length of the edge of the tenth cube (the largest)?

  b If the cubes are placed on top of each other, what is the height of the tower?

15   Two towers are built with the cubes in question *14*. One tower is made of cubes whose edges are an even number of centimetres in length, the other with cubes whose edges are an odd number of centimetres in length.

     Which tower is higher, and by how much? Can you see a quick way to find the answer?

---

3 Area

Which of the shapes in Figure 1 covers more paper?
In other words, which shape has the greater area?

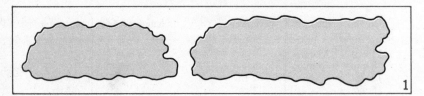

1

One way to find out is to trace each shape on to tracing paper, and then to put your tracing on top of a sheet of squared paper. Count the number of complete squares in the shape, and also count as whole squares fractions of squares which are half or more than half a square, ignoring smaller fractions. This will give you an approximate value of the area in each case.

If possible, try this on 5-mm and 2-mm squared paper, and on isometric graph paper.

Example.—Find the area of the shaded region in Figure 2.

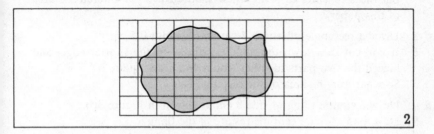

2

The number of complete squares is 13.
The number of half-squares, or more than half-squares, is 11.
So the area consists of approximately 24 squares.

Exercise 4

1   In your geography atlas find the map of Europe, and look at France
and Spain.
    Which appears to be the larger country? Compare their areas by
the method given above.

2   Find the map of Australia. Compare the areas of Queensland and
Western Australia by the same method.

*   *   *   *

Units of measure of area

(i) *The square centimetre.*—The standard practice is to use squares
for comparing areas, and one useful unit is the square centimetre.
    Draw, and colour or shade, a square of side 1 cm.
    Its area is 1 *square centimetre*, or 1 cm² (read as 'one square
centimetre').

Exercise 5

1 a Draw, and shade, a square of side 2 cm.
  b How many squares of side 1 cm does it contain?
  c So what is its area?

2   What would be the area of a square of side:

  a 3 cm   *b* 4 cm   *c* 5 cm   *d* 10 cm?

3   Cut out a square of side 5 cm. Cut this into three or four parts, and put these together to form a new shape. Have you altered the area of this paper?

4 *a*   Draw a rectangle of length 2 cm and breadth 0·5 cm.

   b   If you cut this down the middle, parallel to the shorter edge, and placed the two parts together, could you get a square?

   c   So what was the area of the rectangle?

5 *a*   Draw a square tiling of side 8 cm as shown in Figure 3(i).

   b   How many square centimetres are in the tiling?

   c   Cut the square along the black lines, and make some different shapes—one is shown in Figure 3(ii).

   d   What is the total area of every shape you can make?

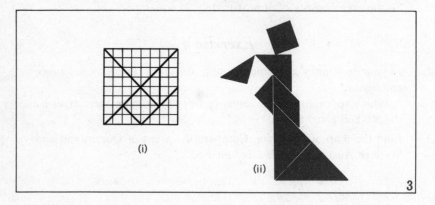

(i)

(ii)

3

(ii) *The square metre.*—The area of a square of side 1 metre is 1 square metre (1 m²).

Draw a square of side 1 m on the classroom floor with chalk, or cut out a paper square of side 1 m.

How many square centimetres are there in 1 square metre?

A football pitch has an area of about 5000 m².

(iii) *The square millimetre.*—Look at a sheet of squared paper with lines ruled on it 1 millimetre apart.

How many square millimetres (mm²) are there in 1 square centimetre?

(iv) *The square kilometre.*—Large areas of land like counties or countries are measured in square kilometres.

How many square metres are there in 1 square kilometre (1 km²)?
We now have:

$$1 \text{ cm}^2 = 10 \times 10 \text{ mm}^2 = 100 \text{ mm}^2$$
$$1 \text{ m}^2 \ = 100 \times 100 \text{ cm}^2 = 10000 \text{ cm}^2$$
$$1 \text{ km}^2 = 1000 \times 1000 \text{ m}^2 = 1000000 \text{ m}^2$$

Exercise 6

1 How many cm² are equal to:

a 3 m²   *b* 2 m² 5000 cm²?

2 How many m² are equal to:

a 50000 cm²   *b* 3 million cm²?

3 How many cm² are equal to:

a 600 mm²   *b* 5000 mm²?

4 How many km² are equal to:

a 4 million m²   *b* 500000 m²?

5 Draw a square of side 4 cm. What is its area?
Explain why the 'area of a square of side 4 cm' is quite different from
an 'area of 4 cm²'.

$$* \qquad * \qquad * \qquad *$$

The area of a rectangle and of a square

1 a Draw a rectangle 5 cm by 3 cm on 1-cm squared paper, or if you
draw it on plain paper draw in all the squares of side 1 cm.

b What is the area of the rectangle in cm²?

c How could you find this answer quickly?

2 What is the area of each of the following rectangles?

a 12 cm by 8 cm   *b* 15 cm by 9 cm
c 40 cm by 25 cm   *d* 5 m by 1½ m

3 What is the area of a square of side:

a 9 cm   *b* 20 cm   *c* 50 cm   *d* 7 m   *e* 10 km?

You will have noticed that:

The area of a rectangle = its length × its breadth
The area of a square   = (its length)²

If l is the number of units of length, b the number of units of breadth, and A the number of units of area, we have:

$$\text{Area of rectangle, } A = l \times b, \quad \text{or} \quad A = lb$$
$$\text{Area of square, } \quad A = l \times l, \quad \text{or} \quad A = l^2$$

Note:—In using these formulae, l and b must be in the same unit of length, and then A will be in the corresponding unit of area.

Example.—Find the area of a rectangle 2 m 75 m long and 40 cm broad.

$$l = 2 \text{ m } 75 \text{ cm} = 275 \text{ cm} \qquad A = lb$$
$$b = 40 \text{ cm} \qquad\qquad \text{Area} = 275 \times 40 \text{ cm}^2$$
$$= 11\,000 \text{ cm}^2, \text{ or } 1{\cdot}1 \text{ m}^2$$

Exercise 7A

1   How many centimetre squares are there in rectangles measuring:

a 13 cm by 9 cm   *b* 50 cm by 25 cm
c 90 cm by 15 cm   *d* 1 m by 1 cm?

2   Calculate the areas of the following rectangles:

	a	b	c	d	e	f
Length	20 cm	14 mm	50 cm	15 km	1 m 20 cm	17 m 20 cm
Breadth	15 cm	12 mm	45 cm	13 km	1 m 10 cm	1 m 3 cm

3   Calculate the areas of squares of side:

a 17 mm   *b* 35 m   *c* 8 km   *d* 25 cm   *e* 22 m

4   Which has the greater area—a rectangle measuring 7 cm by 5 cm, or a rectangle measuring 9 cm by 4 cm?

5   Which floor area of two sheds is the greater—one measuring 9 m by 7 m, or one measuring 11 m by 6 m?

6   Which football pitch is bigger—one measuring 110 m by 48 m, or one measuring 100 m by 53 m?

7   Find the areas of rectangles with these lengths and breadths:

a 3 cm, 2 cm   *b* $3\frac{1}{2}$ cm, 2 cm   *c* 3 cm, $2\frac{1}{2}$ cm

Draw the rectangles in *b* and *c* on 1-cm squared paper.
Divide these into cm squares, or simple fractions of cm squares.
Hence check the answers you obtained by calculation.

8   Find the area of the shaded part in each of the shapes in Figure 4.
     Hint.—It is usually best to look for large rectangles out of which
     smaller rectangles have been cut to give the shaded shape. Hence
     the required area can often be found by subtracting the areas of
     various rectangles.

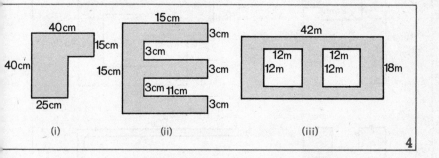

(i)   (ii)   (iii)

4

9   The area of the page of a book is 60 cm². Find the length if the
     breadth is 5 cm.

10   The area of the floor of a room is 90 m². Find the breadth if the
     length is 10 m.

11   The area of a rectangular sheet of glass is 180 cm². Find the breadth
     if the length is 15 cm.

12   The area of a hockey pitch is 3800 m². Find its length if its breadth
     is 50 m.

Exercise 7B

1   Calculate the area of the shaded part of each shape in Figure 5.

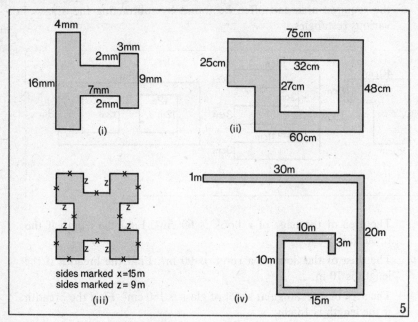

(i)   (ii)   (iii) sides marked x = 15 m, sides marked z = 9 m   (iv)

5

2 *a* On 5-mm squared paper draw a rectangle 12 units long and 6 units broad (use 5-mm units).

 b Show how this area can be covered in two different ways by rectangular tiles each 3 units by 2 units.

 c How many tiles are required in each case?

3   How many rectangular tiles each 5 cm by 3 cm are needed to cover the following rectangular spaces, without leaving any gaps?

 a 40 cm by 21 cm   *b* 36 cm by 25 cm

4   A rectangular hearth 1 m 60 cm long is covered by tiles each 15 cm by 10 cm.

 a How many tiles are there in each row along the length of the hearth?

 b If 48 tiles are used altogether, how wide is the hearth?

5   A rectangular floor 5 m long is covered with square tiles of side 25 cm.

 a How many tiles are there in a strip the length of the room?

 b If 320 tiles are needed altogether, what is the width of the room?

6 Boxes with bases 8 cm long and 6 cm broad are placed in a tray 72 cm long and 48 cm broad so that they all lie the same way, and do not overlap

How many boxes can be put on the tray:

a when the 8-cm side is parallel to the longer side of the tray;

b when the 8-cm side is parallel to the shorter side of the tray?

7 Boxes with bases 8 cm by 10 cm are placed on a tray 75 cm by 55 cm so that they all lie the same way, and do not overlap.

a What is the greatest number of boxes that can be put on the tray in this way?

b Can the boxes be put on the tray in this way without leaving any gaps?

8 An open box is 20 cm long, 15 cm broad, and 10 cm deep. The inside, including the bottom of the box, is lined with aluminium foil. What area of foil is required?

9 A rectangular sheet of plastic has an area of 1 m² 625 cm². If its breadth is 85 cm, calculate its length.

10 A square plot of grass has a side of 5 m, and there is a path 1 m wide round the outside of the plot.

a Calculate the area of the path.

b Find the cost of gravel for the path at a price of 20p per m².

11 What is the total surface area of a cube of edge 5 cm?

12 What is the total surface area of a cuboid 15 cm long, 10 cm broad, and 10 cm deep?

13 Find the perimeters of squares whose areas are:

a 81 cm²   *b* 400 m²

4 Volume

How could you compare the amount of liquid that a jug, a milk bottle, a vase, and a jam jar hold?

We now require a method for measuring the volumes of various objects. Just as we used a square of side 1 cm as a unit to measure areas, we now choose a cube of edge 1 cm as a unit to measure volumes. This has a volume of *1 cubic centimentre*, or 1 cm³ (read as 'one cubic centimetre').

Figure 6(i) shows a cube of side 1 cm, filled with sand. The pile of sand from the cube is shown in Figure 6(ii). Notice that many different shapes can have the same volume.

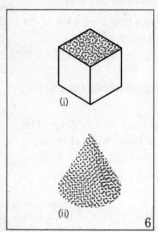

(i)

(ii)

6

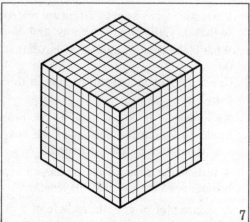

7

Figure 7 shows a cube of side 10 cm, composed of cubes of side 1 cm.

a How many 1-cm cubes are there in one row along the bottom?
b How many such rows are there in the bottom layer?
c So how many 1-cm cubes are there in the bottom layer?
d How many such layers are there in the large cube?
e So how many 1-cm cubes are there in the 10-cm cube?

The volume of this large cube is convenient for measuring everyday liquids such as milk and paraffin, so it has been defined as the unit of volume called *1 litre*.

$$1 \text{ litre} = 1000 \text{ cm}^3$$

Since 1 cm³ is one thousandth part of a litre, it is also called 1 millilitre.

Hence $1 \text{ cm}^3 = 1 \text{ millilitre (1 ml)}$
and we have $1 \text{ litre} = 1000 \text{ cm}^3 = 1000 \text{ ml}$

Common volumes.—You might buy milk in litre or ½-litre bottles, and might see your father ordering 10 or 20 litres of petrol for his car.

The plastic spoon supplied for taking medicine holds 5 ml, and a teacup holds about 150 ml.

Large volumes are usually measured in cubic metres (m³), and very small volumes in cubic millimetres (mm³).

*   *   *   *

The volume of a cuboid

a Sketch a cuboid 4 cm long, 3 cm wide, and 2 cm high. Draw lines to indicate the sides of the 1-cm cubes from which the cuboid could be made.

b How many 1-cm cubes are there in the bottom layer?

c How many such layers are there?

d So how many 1-cm cubes are there in the cuboid?
This gives the volume (or capacity) of the cuboid in cm³.

e How could you find the answer quickly?

You will have noticed that:

The volume of a cuboid = its length × its breadth × its height

If *l*, *b* and *h* are the number of units of length, breadth and height, and *V* the number of units of volume, then:

$$\text{Volume of cuboid, } V = lbh$$

We can write this as $V = (lb)h$. If we notice that lb gives the area, A, of the base, we also have:

$$\text{Volume of cuboid, } V = Ah$$

i.e. Volume of cuboid = Area of base × height

Example 1.—Calculate the volume of a rectangular box 20 cm long, 15 cm broad, and 10 cm high.

$l = 20$ cm	$V = lbh$
$b = 15$ cm	Volume $= 20 \times 15 \times 10$ cm³
$h = 10$ cm	$= 3000$ cm³

Example 2.—Find the volume, in litres, of a cuboid with area of base 1200 cm² and height 20 cm.

$A = 1200$ cm²	$V = Ah$
$h = 20$ cm	Volume $= 1200 \times 20$ cm³
	$= 24000$ cm³
	$= 24$ litres

Note.—Great care must be taken with the units of area and volume, and these should always be stated in the answers to questions. *l*, *b* and *h* should all be expressed in the same unit.

Exercise 8A

In this Exercise, l, b, h, A, V have the meanings explained above.

1 Calculate the volumes of the following cuboids:
a $l = 8$ mm, $b = 5$ mm, $h = 4$ mm
b $l = 12$ cm, $b = 9$ cm, $h = 6$ cm
c $l = 50$ cm, $b = 40$ cm, $h = 10$ cm (Give answer in litres.)
d $A = 36$ m², $h = 12$ m
e $A = 7$ cm², $h = 42$ cm
f $A = 250$ cm², $h = 60$ cm (Give answer in litres.)

2 What is the volume of a box 6 cm by 5 cm by 4 cm?

3 What is the volume of a room 5 m long, 4 m wide, and 3 m high?

4 A paddling pool is 10 m long, 7 m broad, and 25 cm deep. How many cubic metres of water can it hold?

5 A cube is made with an edge 5 cm long. Calculate its volume:
a in cm³   b in ml

6 A small glass tank is 25 cm long, 12 cm broad and 8 cm high. Calculate its volume:
a in cm³   b in ml   c in litres

7 A beaker contains 250 ml of water. How many such beakers filled with water would be needed to fill an empty tank that can hold 3 litres?

8 A rectangular container for orange juice is 50 cm long, 50 cm broad, and 20 cm deep.
a Calculate its volume, in litres.
b If the orange juice is supplied in 10-litre drums, how many of these would be required to fill an empty container?

9 A classroom is 8 m long, 7 m broad, and 3 m high.
a If each pupil is supposed to have 6 m³ of air space, how many pupils can the room hold?
b How much air space is allowed for each pupil in your classroom?

10 A fish tank is 1 m long, 25 cm wide, and 20 cm deep. How many litres of water does it hold?

11 A boy builds 27 cubes, each of edge 1 cm, into a single large cube.
a What is the volume of the large cube?
b What is the length of the edge of the large cube?

c How many more 1-cm cubes would he need to build a cube with side 1 cm longer than the first one?

12 A book has a volume of 480 cm³. Its length is 20 cm, and its breadth is 12 cm. Calculate its thickness.

13 A matchstick is 4 cm long, and the end is 2 mm by 2 mm.

a Calculate the volume of a matchstick in mm³.

b What would be the volume of 50 matchsticks filling a matchbox? Give your answer in mm³, and in cm³.

14 The area of the base of a rectangular paving slab is 2500 cm², and the thickness of the slab is 5 cm.

a Calculate the volume of one slab.

b Calculate also the volume of material in 80 slabs, giving your answer in cm³. Explain why this volume is also 1 m³.

Exercise 8B

In this Exercise, *l*, *b*, *h*, *A*, *V* have the meanings explained earlier in the Section.

1 For each of the following cuboids, give your answer in the unit stated:

a $l = 12$ mm, $b = 8$ mm, $h = 6$ mm. Find V in mm³.

b $l = 2$ m, $b = 50$ cm, $h = 40$ cm. Find V in litres.

c $l = 1$ km, $b = 3$ m, $h = 50$ cm. Find V in m³.

d $l = 25$ cm, $b = 24$ cm, $V = 3$ litres. Find h in cm.

e $b = 6$ cm, $h = 2$ cm, $V = 3$ m³.   Find l in km.

2 Find the volume of each of the following cuboids, giving your answers in the units stated:

a $A = 28$ m², $h = 15$ m; m³

b $A = 2$ cm², $h = 7$ m; cm³

c $A = 1$ cm², $h = 1$ km; m³

3 A tank has a square base of side 40 cm, and is 30 cm deep. If it is empty, and water is run in at the rate of 1 litre per minute, how long will it take for the water:

a to reach a depth of 10 cm;

b to half fill the tank?

Do you need to know the depth of the tank in *a* or *b*?

4   A water tank has a square base of side 60 cm, and a height of 40 cm, and it is full of water. It develops a leak, and loses $\frac{1}{2}$ cm³ of water every second.

a   How many litres are lost in 12 hours?

b   By how much will the level of water fall in this time?

5   Oil is kept in a rectangular tank 40 cm long, 30 cm wide and 80 cm high. Drops from a leak in the tank fall into a tray 30 cm by 20 cm by 3 cm deep.

a   If the level of oil in the tank falls by 1 cm, what volume of oil in the tank has leaked out?

b   By how much does the level of the oil in the tray rise?

c   What information that was given in the question did you not need?

6   A youngster has 100 small cubes, each of edge 1 unit.

a   What is the volume of the largest cube he can make with them, and what is the length of its edge?

b   Answer the same question if 1000 small cubes were available.

c   Answer the same question for 10000 small cubes.

7   Two cubes of edges 5 cm and 6 cm are made from small cubes of side 1 cm. If all the small cubes in the two large ones were put together they would almost make another cube with a whole number of centimetres in its edge.

What would the length of this edge be, and how many more small cubes would be needed to complete the large one?

8   Bricks 5 cm by 3 cm by 2 cm have to be packed in a box 22 cm by 16 cm by 14 cm, so that all the bricks lie the same way. There are six different ways in which this can be done.

a   How many bricks can be packed in each way?

b   When the greatest number of bricks is packed in, what is the volume of the space left in the box?

9   Rain falls to a depth of 1 cm on a playing field of area 20000 m². If water weighs approximately 1 kg per litre, what is the weight in kg of water that falls on the playing field?

5 Mass (or weight)

The basic unit of mass in the metric system is the kilogramme (kg), which is divided into 1000 grammes (g).

Thus $\qquad$ 1 kg = 1000 g

In fact, $\quad$ 1 g is the mass of 1 cm³ of water.

So $\qquad$ 1000 g is the mass of 1000 cm³ of water

i.e. $\qquad$ 1 kg is the mass of 1 litre of water.

Pass a mass of 1 kg, and another of 1 g, round the class.

For scientific purposes, 1 gramme is divided into 1000 milligrammes (mg).

So $\qquad$ 1 g = 1000 mg.

Example 1. $\qquad$ 5780 g = 5000 g + 780 g

$$= 5 \text{ kg} + 780 \text{ g}$$
$$= 5 \cdot 780 \text{ kg}$$

Example 2. $\qquad$ 289 mg = 0 g + 289 mg

$$= 0 \cdot 289 \text{ g}$$

Example 3. $\qquad$ 25 g = 0·025 kg

Remember that 500 g = $\frac{1}{2}$ kg, and 250 g = $\frac{1}{4}$ kg.

Exercise 9

1 $\quad$ Express in kg: 4000 g, 2872 g, 479 g, 75 g

2 $\quad$ Express in g: 2·5 kg, 0·105 kg, 0·003 kg, $\frac{1}{4}$ kg, $\frac{1}{2}$ kg, $\frac{3}{4}$ kg

3 $\quad$ Express in g: 5000 mg, 9810 mg, 500 mg, 25 mg

4 $\quad$ Express in mg: 7 g, 7·5 g, 7·52 g, 0·075 g

5 $\quad$ Calculate the total mass of 2 packets of cereal each of mass 450 g, 4 tins of fruit each of mass 125 g, and $\frac{1}{2}$ kg of butter.

6 $\quad$ A bar of chocolate has mass 125 g, and costs 11p. Calculate the total mass and cost of 25 bars of chocolate.

7 $\quad$ A school shop sells 850 packets of potato crisps in a month. Each packet contains 22 g of crisps and costs $2\frac{1}{2}$p.

$\qquad$ Calculate the number of kg of crisps sold and the total money received.

8 A box contains 36 tins of cleaning powder. Each tin contains 225 g of the powder, and the tin itself has a mass of 15 g.

    If the full box has a mass of 10 kg, what is the mass of the empty box?

9 Find the total mass in kg of a Christmas parcel containing:

    6 packets of tea, each 125 g

    12 bags of sugar, each 500 g

    3 Christmas cakes, each 750 g

    10 tins of meat, each 175 g

10 A $3\frac{1}{4}$ kg piece of steak is bought to supply a meal for 25 people.

  a How many grammes should each person get?

  b If the steak costs 96p per kg, what is the total cost of the steak, and the approximate cost per person (to the nearest halfpenny)?

11 A rectangular water tank has a square base of side 30 cm and is 1 m high. The empty tank has a mass of 12·5 kg, and 1 litre of water has a mass of 1 kg.

    Find the total mass of the tank when full of water.

12 An aeroplane has 16 first-class and 28 tourist-class seats. First-class passengers are allowed 44 kg of luggage each and tourist-class passengers 30 kg each.

    Assuming that the average mass of each passenger is 85 kg, what is the pay-load of the plane? (The pay-load is the total mass of passengers and luggage when all seats are filled and every passenger takes his full allowance of luggage.)

Summary

Length
$1 \text{ cm} = 10 \text{ mm}$
$1 \text{ m} = 100 \text{ cm} = 1000 \text{ mm}$
$1 \text{ km} = 1000 \text{ m}$

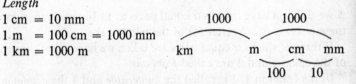

Area
$1 \text{ cm}^2 = 100 \text{ mm}^2$
$1 \text{ m}^2 = 10000 \text{ cm}^2$ *For a rectangle, $A = lb$*
$1 \text{ km}^2 = 1000000 \text{ m}^2$ *For a square, $A = l^2$*

Volume
$1 \text{ litre} = 1000 \text{ ml} = 1000 \text{ cm}^3$
$1 \text{ m}^3 = 1000000 \text{ cm}^3 = 1000 \text{ litres}$

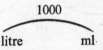

For a cuboid, $V = lbh$, or $V = Ah$

Mass
$1 \text{ g} = 1000 \text{ mg}$
$1 \text{ kg} = 1000 \text{ g}$

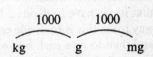

Fractions, Ratios and Percentages

1 The meaning of a fraction

If we divide a cake into four equal parts as in Figure 1, then each of these parts is one quarter ($\frac{1}{4}$) of the whole cake.

If three of the four equal parts are taken we have *three quarters* ($\frac{3}{4}$) of the cake. $\frac{1}{4}$ and $\frac{3}{4}$ are called *fractions*.

In the fraction $\frac{1}{4}$, 1 is called the *numerator* and 4 the *denominator*. What are the numerator and denominator of the fraction $\frac{3}{4}$?

 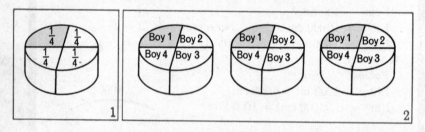

The fraction $\frac{3}{4}$ can arise in another way. Suppose we have three identical cakes which are to be divided equally among four boys; that is, we have to divide 3 by 4 to get each boy's share. The easiest way to do this is to divide each of the three cakes into four quarters, and then to give each boy one quarter from each cake, as shown in Figure 2. Each boy's total share therefore is the same as if he had received three quarters ($\frac{3}{4}$) of one whole cake and so we see that

$$3 \div 4 = \tfrac{3}{4} \; (= \text{three quarters}).$$

What fraction of a whole apple does each boy get if 2 apples are divided among 3 boys? 5 apples among 8 boys? 3 apples among 5 boys?

You should now realize that the same fraction can arise in two different ways,

e.g. $\qquad\qquad \tfrac{2}{3} = \text{two thirds of one,} \quad and \quad \tfrac{2}{3} = 2 \div 3$

Exercise 1

1   Draw a rectangle 8 cm by 5 cm, and divide it up as shown in Figure 3.

First strip	1
Second strip	$\frac{1}{2}$
Third strip	$\frac{1}{4}$
Fourth strip	$\frac{1}{8}$
Fifth strip	$\frac{1}{16}$

3

Taking each of the five equal strips to represent 1, shade in areas corresponding to the fractions:

a $\frac{1}{2}$ *b* $\frac{3}{4}$ *c* $\frac{5}{8}$ *d* $\frac{11}{16}$ *e* $\frac{1}{32}$.

2   If a straight line measures 12 cm, how long would one quarter of the line be? one third? one half? one sixth? one eighth? two-thirds? five-sixths? one tenth?

3   What fraction of a calendar week (of seven days) is a normal school week?

4   What fraction of the months of the year begin with N? J? M?

5   What fraction of the months of the year have exactly 30 days?

6   What fraction of the letters of the alphabet are vowels? consonants?

7   Express the number of mathematics periods in your weekly time-table as a fraction of the total number of periods.

8   Express the length of your arithmetic period as a fraction of one hour.

9   What fraction of $\frac{1}{4}$ hour is 1 minute? 7 minutes? 10 minutes?

10   What fraction of one complete turn is a right angle? a straight angle? three right angles?

11   What fraction of 1 hour is 1 minute? 30 minutes? 1 second? 60 minutes?

12*a* Write down the numerators of the fractions $\frac{2}{3}$, $\frac{5}{7}$, $\frac{1}{2}$, $\frac{9}{11}$.

b Write down the denominators of the fractions $\frac{3}{5}$, $\frac{1}{4}$, $\frac{7}{8}$, $\frac{99}{100}$.

2 Equal fractions

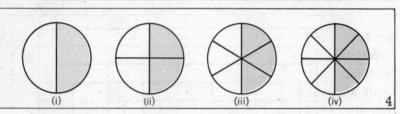

(i)   (ii)   (iii)   (iv)   4

What fraction of the circle is shaded in each diagram?

Now give the answer for diagram (ii) in quarters, (iii) in sixths, (iv) in eighths.

Write down the four equal fractions you have now obtained. Which one is in the simplest form?

Look back at Figure 3, and write down two fractions which are equal to $\frac{3}{4}$.

You should have found that:

$$\frac{1}{2} = \frac{2}{4} = \frac{3}{6} = \frac{4}{8}, \quad \text{and that} \quad \frac{3}{4} = \frac{6}{8} = \frac{12}{16}.$$

Notice that we can obtain sets of equal fractions by multiplying, or dividing, the numerator and denominator by the same number. For example,

$$\frac{1}{2} = \frac{1 \times 3}{2 \times 3} = \frac{3}{6}, \quad \text{and} \quad \frac{12}{16} = \frac{12 \div 4}{16 \div 4} = \frac{3}{4}.$$

This section can be summed up by saying that all the fractions $\frac{3}{4}$, $\frac{6}{8}$, $\frac{9}{12}$, $\frac{12}{16}$, ..., $\frac{30}{40}$, ..., $\frac{75}{100}$, ..., $\frac{3000}{4000}$, ... are equal to the fraction $\frac{3}{4}$. $\frac{3}{4}$ is the simplest form of all of these fractions.

$\frac{3}{4}$, $\frac{6}{8}$, $\frac{9}{12}$, etc., are sometimes called *equivalent fractions*.

We already know that $\frac{75}{100}$ can be written as 0·75, which is the decimal form of the fraction $\frac{3}{4}$.

Exercise 2A

1 a Draw a rectangle. Divide it up so that you can shade one half of it.

 b Sketch it again and try to divide it in a different way to show one half again.

 c A rectangular block of butter can be halved in several different ways. Describe some of these ways. Which is easiest?

2   Draw a square and divide it up in such a way that you can shade
    one quarter of it. Try to do this in different ways.
    Give two equivalent fractions for each of the following:

3   $\frac{1}{5}$ 4 $\frac{2}{3}$ 5 $\frac{5}{9}$ 6 $\frac{16}{24}$

7   Fill in the missing numerators:

$$\frac{2}{3} = \frac{}{9} = \frac{}{12} = \frac{}{18} = \frac{}{30} = \frac{}{150}$$

    Find the simplest form of each of the following fractions:

8 a $\frac{2}{6}$ b $\frac{4}{6}$ 9 a $\frac{8}{12}$ b $\frac{9}{12}$ 10a $\frac{5}{10}$ b $\frac{8}{10}$

11a $\frac{9}{15}$ b $\frac{12}{15}$ 12a $\frac{6}{10}$ b $\frac{10}{10}$ 13a $\frac{14}{21}$ b $\frac{18}{21}$

14a $\frac{6}{12}$ b $\frac{12}{12}$ 15a $\frac{6}{18}$ b $\frac{12}{18}$ 16 $\frac{11}{33}$

17 $\frac{20}{25}$ 18 $\frac{13}{26}$ 19 $\frac{28}{32}$

20 Explain your method of simplifying question 19.

21 Each circle in Figure 5 is divided into equal parts. What fraction is
    the shaded area of the whole circle in each case?

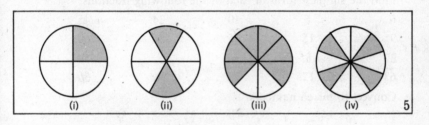

(i)   (ii)   (iii)   (iv)   5

            *   *   *   *

$\frac{7}{2} = \frac{6}{2} + \frac{1}{2} = 3 + \frac{1}{2} = 3\frac{1}{2}$,   and $2\frac{3}{4} = 2 + \frac{3}{4} = \frac{8}{4} + \frac{3}{4} = \frac{11}{4}$.

$3\frac{1}{2}$ and $2\frac{3}{4}$ are called *mixed numbers* (i.e. each is a whole number and
a fraction).

    Convert each of the following fractions into a mixed number
    (i.e. a whole number and a fraction):

22 $\frac{9}{5}$ 23 $\frac{11}{3}$ 24 $\frac{15}{4}$ 25 $\frac{23}{4}$

26 $\frac{27}{5}$ 27 $\frac{55}{9}$ 28 $\frac{75}{11}$ 29 $\frac{21}{12}$

    Convert each of the following into fractional form:

30 $2\frac{3}{4}$ 31 $5\frac{1}{2}$ 32 $7\frac{2}{3}$ 33 $3\frac{3}{5}$

34 $10\frac{1}{10}$ 35 $1\frac{1}{8}$ 36 $6\frac{1}{4}$ 37 $11\frac{5}{12}$

    Give two equal fractions for each of the following:

38 $\frac{1}{2}$ 39 $\frac{15}{25}$ 40 $\frac{3}{4}$ 41 $\frac{19}{20}$

Exercise 2B

1   Draw a circle, with two perpendicular diameters. Lightly shade in three-quarters of the circle. Now draw two more diameters to show that $\frac{3}{4} = \frac{6}{8}$.

2 a   Construct a diagram like Figure 3 to show the relationship between thirds, sixths, and twelfths.

  b   How many twelfths equal one third? two thirds? five sixths?

3 a   Complete the following:
$$\frac{7}{12} = \frac{14}{-} = \frac{35}{-} = \frac{-}{36} = \frac{-}{120} = \frac{7x}{-}$$

  b   List the numerators of all the fractions in *a*

4   What is the simplest fraction equal to each of the fractions $\frac{15}{18}, \frac{20}{24}, \frac{65}{78}, \ldots$?

Find the simplest form of each of the following fractions:

5   $\dfrac{6}{9}$ 6 $\dfrac{6}{15}$ 7 $\dfrac{10}{25}$ 8 $\dfrac{24}{36}$ 9 $\dfrac{45}{60}$

10   $\dfrac{27}{63}$ 11 $\dfrac{85}{125}$ 12 $\dfrac{5a}{7a}$ 13 $\dfrac{2x}{3x}$ 14 $\dfrac{5a}{5b}$

Convert to mixed numbers:

15   $\frac{8}{5}$ 16 $\frac{16}{5}$ 17 $\frac{29}{8}$ 18 $\frac{31}{12}$ 19 $\frac{59}{9}$ 20 $\frac{100}{3}$

Convert to fractional form:

21   $1\frac{1}{4}$ 22 $5\frac{1}{3}$ 23 $9\frac{5}{8}$ 24 $16\frac{7}{8}$ 25 $41\frac{2}{3}$

Give two equivalent fractions for each of the following:

26   $\dfrac{2}{5}$ 27 $\dfrac{10}{32}$ 28 $\dfrac{4}{7}$ 29 $\dfrac{35}{100}$ 30 $\dfrac{x}{y}$

31   0·5   32   0·6   33   0·25   34 $\dfrac{x}{2}$ 35 $\dfrac{2}{x}$

36   Which two of the following fractions are equal?
$$\frac{45}{65}, \quad \frac{75}{105}, \quad \frac{63}{90}, \quad \frac{54}{78}$$

3 From whole numbers to fractions

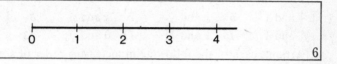

6

We can show the whole numbers in order on a *number line* by marking points equal distances apart, and putting the numbers opposite the marks as shown.

Imagine that the diagram is a road with milestones showing the distances from the starting point at the left end of the road. Suppose we now start from this end and walk half a mile along the road, stop for a rest, then walk half a mile farther on, and so on. We could show our stopping places like this.

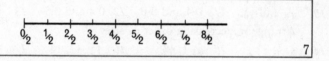

7

Which of these fractions are equal to the whole numbers 0, 1, 2, 3, 4?

Suppose we start at the beginning again and run for a quarter of a mile, then rest, then run for another quarter of a mile, and so on. Draw a diagram to show the stopping places this time, and list the set of whole numbers in the form obtained here.

Do you see that we could proceed to divide up the road, and hence the number line, into smaller and smaller parts? In fact every fraction you can think of can be represented by a point on the number line.

Notice that (i) $2 = \frac{2}{1} = \frac{4}{2} = \frac{6}{3} = \frac{8}{4} = \ldots$

(ii) We can always find as many fractions between two given fractions as we wish.

For example, a fraction between $\frac{1}{3}$ and $\frac{1}{2}$, i.e. $\frac{2}{6}$ and $\frac{3}{6}$, i.e. $\frac{4}{12}$ and $\frac{6}{12}$, is $\frac{5}{12}$.

Exercise 3A

Write each of the following as fractions in two different ways:

1 1   *2* 10   *3* 25   *4* $\frac{1}{2}$

Find one fraction between:

5 $\frac{1}{2}$ and 1   *6* $\frac{1}{4}$ and $\frac{3}{4}$ *7* $\frac{1}{2}$ and $\frac{5}{6}$ *8* $\frac{1}{4}$ and $\frac{1}{2}$

9 $\frac{1}{5}$ and $\frac{1}{2}$ *10* 1 and $1\frac{1}{8}$ *11* 0 and $\frac{1}{10}$ *12* $\frac{1}{4}$ and $\frac{1}{3}$

Arrange each of the following in increasing order of size:

13 $\frac{1}{2}, \frac{1}{4}, \frac{3}{4}$ *14* $\frac{1}{2}, \frac{5}{8}, \frac{3}{8}$ *15* $\frac{2}{3}, \frac{1}{2}, \frac{3}{4}$ *16* $\frac{1}{2}, \frac{2}{5}, \frac{3}{5}$

17 Insert two fractions between *a* $\frac{1}{2}$ and $\frac{1}{3}$ *b* $\frac{1}{5}$ and $\frac{1}{10}$.

18 Insert three fractions between $\frac{1}{2}$ and 1.

Exercise 3B

Write each of the following as fractions in three different ways:

1 3   *2* 15   *3* 100   *4* $\frac{3}{4}$ *5* x

Find one fraction between:

6 $\frac{1}{3}$ and 1   *7* $\frac{1}{3}$ and $\frac{2}{3}$ *8* $\frac{1}{4}$ and $\frac{1}{5}$ *9* 0 and $\frac{1}{8}$

10 $\frac{1}{10}$ and $\frac{1}{100}$ *11* 0.4 and 0.6   *12* 0.4 and 0.5   *13* $\frac{2}{3}$ and 1

Arrange in decreasing order of size:

14 $\frac{1}{2}, \frac{1}{8}, \frac{1}{4}$ *15* $\frac{7}{8}, 1, \frac{11}{12}$ *16* $1\frac{2}{3}, 1\frac{3}{4}, 1\frac{5}{6}$ *17* $\frac{4}{5}, \frac{7}{10}, \frac{69}{100}$

18 Which of the following fractions lie between $\frac{3}{4}$ and $\frac{7}{8}$?

$$\frac{1}{2}, \frac{5}{6}, \frac{19}{24}, \frac{2}{3}$$

19 Insert nine fractions between $\frac{1}{5}$ and $\frac{1}{6}$.

*   *   *   *

Using fractions

Example 1. $\frac{2}{5}$ of £3·50 = 2 × £0·70 = £1·40.

Example 2. What fraction of a right angle is 15°?

Since 1 right angle = 90°, the fraction = $\frac{15}{90} = \frac{3}{18} = \frac{1}{6}$.

Example 3. It takes 4 hours to travel $\frac{2}{3}$ of a journey at a constant speed. How long will the whole journey take at the same speed?

$\frac{2}{3}$ of the journey takes 4 hours

so $\frac{1}{3}$ of the journey takes 4 hours ÷ 2 = 2 hours

so $\frac{3}{3}$ of the journey takes 2 hours × 3 = 6 hours

i.e.   the whole journey takes 6 hours

Reminders. 1 kg = 1000 g; 1 litre = 1000 ml; 1 m = 100 cm

Exercise 4A

Work out the values in questions *1–9*:

1 $\frac{1}{3}$ of £4·74   2 $\frac{2}{3}$ of £1·05   3 $\frac{5}{9}$ of 63p

4 $\frac{3}{4}$ of 1 litre (in ml) 5 $\frac{5}{6}$ of 1 hour   6 $\frac{7}{10}$ of 1 metre

7 $\frac{1}{4}$ of 1 kg (in g)   8 $\frac{3}{4}$ of 1 kg   9 $\frac{7}{8}$ of £5·68

10 What fraction of a right angle is 1°? 20°? 85°? $\frac{1}{2}$°?

11 What fraction of:

  a 1 kg is 800 g   b 1 litre is 200 ml   c 1 m is 64 cm?

12 A girl's pocket money is £1·60 per month. How much would it be if it was increased by:

  a $\frac{1}{10}$ b $\frac{1}{8}$ c $\frac{1}{5}$ d $\frac{1}{4}$?

13 How many degrees are there in $\frac{2}{3}$ of a right angle? In $\frac{1}{4}$ of a right angle? In $1\frac{1}{2}$ right angles?

14 A man's weekly wage is £20. After a year it is increased by $\frac{3}{40}$. What is the wage then?

15 18-carat gold contains $\frac{18}{24}$ pure gold, and the rest alloy.

  a How much pure gold is there in 144 g of 18-carat gold?
  b How much pure gold is there in 312 g of 10-carat gold?

16 Calculate the fraction of each day (24 hours) that you spend in school.

17 If it takes half an hour to travel $\frac{3}{8}$ of a journey how long will it take for the whole journey, assuming the same average speed?

18 A tank was $\frac{3}{4}$ full of water when it contained 276 litres. How many litres would it contain when full? $\frac{1}{2}$ full? $\frac{5}{8}$ full?

19 An aircraft uses $\frac{3}{5}$ of its fuel in flying 1230 km. How far can it fly on its remaining fuel?

20 An office block is to be 20 storeys high, all the storeys being of the same height. The 6 storeys completed reach a height of 33 metres.

  a What fraction has been built?
  b What will the total height be?

Exercise 4B

Evaluate each of the following, in questions *1–6*:

1 $\frac{3}{7}$ of 98 pence   2 $\frac{4}{5}$ of £3·85   3 $\frac{5}{12}$ of 1 hour

4 $\frac{7}{10}$ of 1 metre   *5* $\frac{3}{8}$ of 2 hours   *6* $\frac{4}{9}$ of 126 kg

7   Calculate in degrees:
 a $\frac{1}{2}$ of a right angle   *b* $\frac{3}{4}$ of a right angle
 c $1\frac{2}{3}$ right angles   *d* $3\frac{1}{2}$ right angles.

8   What fraction of a right angle is 10°? 35°? 2°? $11\frac{1}{4}$°? $\frac{3}{4}$°?

9   What angle (as a fraction of a right angle) must be added to $\frac{1}{4}$ right angle to make up *a* 1 right angle *b* 2 right angles *c* 4 right angles?

10 Repeat question *9* for $\frac{2}{3}$ right angle, and for $\frac{3}{8}$ right angle.

11 On his birthday a boy had his pocket money increased from 25p to 30p.
      Express the increase as a fraction of *a* his old pocket money *b* his new pocket money.

12 The daily sales in a shop totalled £126·25, £243·54, £98·98, £229·83, £147·07, £207·13 during one week.
 a If one quarter of the sales represent profit, calculate the total profit.
 b What was the cost price to the shopkeeper of all the goods sold that week?

13 I started painting a fence at 1 p.m. By 7 p.m. I had painted $\frac{3}{4}$ of it, and I stopped for half an hour to have a meal.
      Working at the same rate, would I be able to finish the painting before darkness fell at 10 p.m.?

14 A man earns £24 per week. He spends $\frac{3}{16}$ of this on rent and he saves £2·70.
      How much has he left to spend weekly on other things? What fraction is this of his income?

15 The petrol tank of a car can hold 30 litres and its petrol consumption is 16 km per litre. Before starting a journey the tank is $\frac{4}{5}$ full:
 a What fraction of the tank still contains petrol after a journey of 192 km?
 b What fraction of the original contents has been used for the journey?

4 The operations of addition, subtraction, multiplication, and division

Addition and subtraction

Example. $\frac{3}{4}+\frac{5}{6}$

The set of multiples of 4 = {0, 4, 8, 12, 16, 20, 24, 28, ...}

The set of multiples of 6 = {0, 6, 12, 18, 24, 30, ...}

So the set of common multiples of 4 and 6 = {0, 12, 24, ...}

Now $\frac{3}{4} = \frac{9}{12} = \frac{18}{24} = ...$, and $\frac{5}{6} = \frac{10}{12} = \frac{20}{24} = ...$

It follows that $\frac{3}{4}+\frac{5}{6}$

$\quad = \frac{9}{12}+\frac{10}{12}$

$\quad = \frac{19}{12}$

$\quad = 1\frac{7}{12}$

Also $\frac{3}{4}+\frac{5}{6}$

$\quad = \frac{18}{24}+\frac{20}{24}$

$\quad = \frac{38}{24}$

$\quad = \frac{19}{12}$

$\quad = 1\frac{7}{12}$

The idea in adding the two fractions is to use equivalent fractions with the same denominator, a *common denominator*.

The working is easier if we use 12, the LCM of 4 and 6, as the common denominator here.

To add or subtract fractions, add or subtract equivalent fractions with a common denominator which is the LCM of the given denominators.

Example 1. $2\frac{3}{8}+1\frac{1}{6} = 2\frac{9}{24}+1\frac{4}{24}$ (24 is the LCM of 8 and 6)

$\quad = 3\frac{9+4}{24}$

$\quad = 3\frac{13}{24}$

Example 2. $3\frac{5}{6}-1\frac{1}{9} = 3\frac{15}{18}-1\frac{2}{18}$ (18 is the LCM of 6 and 9)

$\quad = 2\frac{15-2}{18}$

$\quad = 2\frac{13}{18}$

Example 3. $3\frac{5}{6}-1\frac{8}{9} = 3\frac{15}{18}-1\frac{16}{18}$

$\quad = 2\frac{15-16}{18}$

$\quad = 1\frac{18+15-16}{18}$ $\quad$ ($2 = 1+1 = 1+\frac{18}{18}$)

$\quad = 1\frac{17}{18}$

The operation involved in adding fractions is based on that of adding whole numbers (9+4 in Example 1).

Do you remember the commutative law for the addition of whole numbers? $a+b = b+a$.

And the associative law? $(a+b)+c = a+(b+c)$.

We now assume that these are true for a, b, c, representing fractions. (And you can verify them by illustrating some examples on the number line.)

For example,
$$\frac{3}{4}+2+1\frac{1}{4}$$
$$= \frac{3}{4}+(2+1\frac{1}{4})$$
$$= \frac{3}{4}+(1\frac{1}{4}+2)$$
$$= (\frac{3}{4}+1\frac{1}{4})+2$$
$$= 2+2$$
$$= 4$$

Exercise 5A

Give the LCM in each of the following:

1	2, 3	**2**	6, 8	**3**	5, 7	**4**	8, 12
5	8, 9	**6**	5, 15, 30	**7**	4, 6	**8**	4, 5, 8

Simplify the following:

9	$\frac{1}{2}+\frac{1}{3}$	**10**	$\frac{1}{3}+\frac{2}{5}$	**11**	$\frac{5}{6}-\frac{1}{8}$	**12**	$\frac{2}{3}+\frac{3}{4}$
13	$\frac{5}{12}+\frac{1}{3}$	**14**	$\frac{2}{3}+\frac{5}{12}$	**15**	$\frac{5}{6}-\frac{3}{4}$	**16**	$1\frac{1}{3}+2\frac{1}{5}$
17	$3\frac{1}{4}+2\frac{3}{5}$	**18**	$3\frac{3}{5}-2\frac{1}{4}$	**19**	$7\frac{1}{2}-2\frac{1}{7}$	**20**	$4\frac{1}{3}+1\frac{3}{4}$
21	$4\frac{2}{5}+3\frac{3}{4}$	**22**	$2\frac{1}{2}-\frac{3}{8}$	**23**	$5\frac{1}{2}+6\frac{1}{5}$	**24**	$1\frac{1}{2}-\frac{1}{5}$

25 Explain in words how you worked out question **9**.

26 In a garden $\frac{1}{3}$ of the area is used for vegetables and $\frac{2}{5}$ for flowers and paths. If the rest of the garden consists of lawns, what fraction of the garden is given to lawns?

27 $\frac{5}{12}$ of the seating capacity of a cinema is in the 'back' stalls, $\frac{1}{4}$ in the 'front' stalls, and the remainder in the balcony.

If the cinema can seat 720 persons, how many can sit in the balcony?

28 Copy and complete these magic squares, in both of which each row, column, and diagonal has the same total:

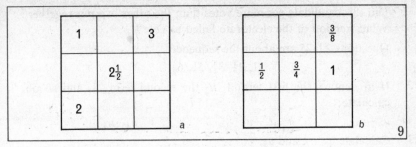

29 Simplify: *a* $3\frac{1}{2}-1\frac{3}{4}$ *b* $4\frac{1}{3}-2\frac{1}{2}$ *c* $6\frac{2}{5}-2\frac{7}{10}$ *d* $1\frac{1}{10}-\frac{1}{5}$

Exercise 5B

Simplify the following:

1 $\frac{7}{8}+\frac{5}{12}$	*2* $1\frac{1}{2}-\frac{3}{8}$	*3* $2\frac{1}{2}+1\frac{7}{12}$	*4* $\frac{1}{2}+\frac{1}{3}+\frac{1}{4}$
5 $1\frac{3}{4}-\frac{5}{6}+2\frac{1}{2}$	*6* $5-1\frac{1}{2}-\frac{5}{8}$	*7* $3\frac{1}{4}-2\frac{3}{4}+2\frac{1}{2}$	*8* $6\frac{1}{2}+2\frac{2}{3}-1\frac{4}{5}$

9 Which is greater, $\frac{13}{24}$ or $\frac{17}{32}$? By how much?

10 Find the sum of $1\frac{1}{2}$ and $2\frac{3}{8}$.

11 Find the difference between $1\frac{11}{16}$ and $2\frac{1}{2}$.

12 By how much is the sum of $1\frac{2}{3}$ and $3\frac{1}{4}$ greater than 4?

13 By how much is the difference between $\frac{11}{16}$ and $\frac{5}{12}$ less than 1?

Use the commutative and associative laws to help you to work out each of the following in the easiest way:

14 $2\frac{3}{4}+9\frac{1}{2}+1\frac{3}{4}$ **15** $2\frac{1}{3}-\frac{3}{4}+1\frac{2}{3}$

16 $5\frac{3}{5}+6\frac{4}{5}-5\frac{1}{5}$ **17** $7\frac{1}{2}+8\frac{1}{4}-9\frac{3}{4}$

18 If $p=1\frac{1}{2}$, $q=2\frac{2}{3}$, $r=3\frac{3}{4}$ evaluate:

a $p+q+r$ *b* $r+q-p$ *c* $2p+3q+4r$

19 One boy does $\frac{1}{2}$ the weeding in a garden and a second boy does $\frac{1}{3}$ of it. How much remains to be done?

20 A owns $\frac{5}{12}$ of some property, B owns $\frac{7}{16}$, and C the remainder.

a What fraction does C own?

b If C's share is worth £840 find the value of the property, and hence calculate A's share and B's share.

21 At an election the three candidates, A, B, and C polled $\frac{1}{3}$, $\frac{1}{4}$, and $\frac{1}{5}$ of the possible number of votes respectively.

a Which candidate was elected?

b Did this candidate get more votes than the other two put together?

c What fraction of the electorate failed to vote?

Questions *22–25* are about the sequence

$$1, 2\tfrac{3}{8}, 3\tfrac{3}{4}, 5\tfrac{1}{8}, 6\tfrac{1}{2}, \ldots$$

22 If u_1 denotes the first term 1, u_2 the second term $2\tfrac{3}{8}$, and so on, calculate:

a u_2-u_1 *b* u_3-u_2 *c* u_4-u_3.

23 Calculate u_6, u_7, and u_8.

24 What did you notice about your answers to question *22*? Now write down the values of:

a u_7-u_6 *b* $u_{98}-u_{97}$ *c* $u_{r+1}-u_r$

25 Evaluate   *a* u_3-u_1 *b* $u_{10}-u_8$ *c* $u_{r+2}-u_r$

Multiplication

In Figure 10, the top and bottom sides of the square have each been divided into two equal parts, and the left and right sides into three equal parts. So the square has been divided into six equal rectangles.

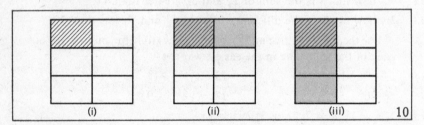

(i)   (ii)   (iii)   10

a If each side of the square is 1 unit long, what is the area of the square?

So, what is the area of each of the six rectangles?

b Look at the red shaded rectangle in Figure 10(i).

What is its length (as a fraction of 1 unit)? What is its breadth? So what is its area?

*From **a** and **b**, you should see that* $\tfrac{1}{2}\times\tfrac{1}{3} = \tfrac{1}{3}\times\tfrac{1}{2} = \tfrac{1}{6}$ I

c In Figure 10(ii), what fraction is the red shaded rectangle of the black shaded rectangle?

What fraction is the black shaded rectangle of the whole square?

It follows that the red rectangle = $\frac{1}{2}$ of $\frac{1}{3}$ of the square.

d In Figure 10(iii), what fraction is the red rectangle of the black rectangle?

What fraction is the black rectangle of the whole square?

It follows that the red rectangle = $\frac{1}{3}$ of $\frac{1}{2}$ of the square.

Also from **a**, the red rectangle = $\frac{1}{6}$ of the square.

*From **c** and **d**, you should see that* $\frac{1}{2}$ of $\frac{1}{3}$ = $\frac{1}{3}$ of $\frac{1}{2}$ = $\frac{1}{6}$ II

From results I and II it is reasonable to write:

$$\frac{1}{2} \text{ of } \frac{1}{3} = \frac{1}{3} \text{ of } \frac{1}{2} = \frac{1}{2} \times \frac{1}{3} = \frac{1}{3} \times \frac{1}{2} = \frac{1}{6} \quad .... \text{ III}$$

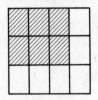

11

Using Figure 11, think out as above why:

$$\frac{3}{4} \text{ of } \frac{2}{3} = \frac{2}{3} \text{ of } \frac{3}{4} = \frac{3}{4} \times \frac{2}{3} = \frac{2}{3} \times \frac{3}{4} = \frac{6}{12} = \frac{1}{2} \quad .... \text{ IV}$$

Notice that in II, $\dfrac{1}{2} \times \dfrac{1}{3} = \dfrac{1 \times 1}{2 \times 3} = \dfrac{1}{6}$,

and in IV, $\dfrac{3}{4} \times \dfrac{2}{3} = \dfrac{3 \times 2}{4 \times 3} = \dfrac{6}{12} = \dfrac{1}{2}$.

To multiply fractions, multiply the numerators together, and the denominators together.

Example 1. $\dfrac{2}{5} \times \dfrac{1}{3} = \dfrac{2 \times 1}{5 \times 3} = \dfrac{2}{15}$

Example 2. $\frac{3}{4}$ of $5 = \frac{3}{4} \times \frac{5}{1} = \frac{15}{4} = 3\frac{3}{4}$

Example 3. $1\frac{1}{3} \times 1\frac{7}{8}$

Method 1	*Method 2*

$$1\frac{1}{3} \times 1\frac{7}{8} = \frac{4}{3} \times \frac{15}{8} \qquad\qquad 1\frac{1}{3} \times 1\frac{7}{8} = \frac{\overset{1}{4}}{\underset{1}{3}} \times \frac{\overset{5}{15}}{\underset{2}{8}}$$

$$= \frac{60}{24} \qquad\qquad\qquad\qquad\qquad = \frac{5}{2}$$

$$= 2\frac{12}{24} \qquad\qquad\qquad\qquad\qquad = 2\frac{1}{2}$$

$$= 2\frac{1}{2}$$

For fractions, as for whole numbers, the operation of multiplication is *commutative* i.e. $a \times b = b \times a$
and *associative* i.e. $(a \times b) \times c = a \times (b \times c)$
The *distributive law* $(a \times c) + (b \times c) = (a+b) \times c$
is also true.

Example 4. $(2\frac{1}{4} \times 1\frac{1}{3}) + (2\frac{3}{4} \times 1\frac{1}{3}) = (2\frac{1}{4} + 2\frac{3}{4}) \times 1\frac{1}{3}$
$$= 5 \times 1\frac{1}{3}$$
$$= 6\frac{2}{3}$$

Exercise 6A

1 a Multiply $\frac{2}{3}$ by 1, by 0, by 3, by 4, by 6, by 18.
  b Multiply $\frac{5}{6}$ by 6, by 2, by 24, by 0, by 3, by 5.
  c Multiply $2\frac{1}{2}$ by 4, by 7, by 0, by 14, by 3.

Simplify each of the following:

2	$\frac{3}{4} \times \frac{2}{9}$	**3**	$\frac{1}{2}$ of $\frac{3}{4}$	**4**	$3\frac{1}{3} \times 1\frac{1}{2}$	**5**	$2\frac{1}{4} \times 1\frac{1}{3}$
6	$\frac{5}{8}$ of $1\frac{1}{5}$	**7**	$2\frac{2}{3} \times 2\frac{1}{4}$	**8**	$6\frac{3}{4} \times \frac{2}{3}$	**9**	$5\frac{1}{4} \times 0$
10	$\frac{1}{4}$ of 8	**11**	$\frac{1}{4} \times 8$	**12**	$\frac{2}{3}$ of $4\frac{1}{2}$	**13**	$\frac{2}{3} \times 4\frac{1}{2}$

14 Find the perimeter of the rectangle whose length is $8\frac{1}{2}$ cm and whose breadth is $6\frac{1}{2}$ cm.

15 Repeat question *14* for the rectangle $5\frac{3}{4}$ cm by $2\frac{1}{4}$ cm.

16 Find the areas of the rectangles in questions *14* and *15*.

17 Calculate the perimeter and the area of the square with side $4\frac{1}{2}$ cm long.

18 Calculate the perimeter and the area of the square with side $1\frac{1}{4}$ m long.

Exercise 6B

1 a Multiply $\frac{3}{8}$ by 1, by 0, by 8, by 2, by 3, by 16.
  b Multiply $1\frac{1}{4}$ by 4, by 2, by 0, by 6, by 8, by 36.
  c Multiply $4\frac{2}{3}$ by 6, by 2, by 12, by 0, by $\frac{1}{2}$, by $1\frac{1}{2}$.

Simplify each of the following:

2	$\frac{2}{3} \times \frac{3}{4}$	**3**	$\frac{3}{4}$ of $\frac{2}{3}$	**4**	$\frac{5}{8} \times 1\frac{1}{3}$	**5**	$2\frac{1}{2} \times 1\frac{2}{3}$
6	$9\frac{3}{4} \times 2\frac{2}{3}$	**7**	$\frac{1}{2}$ of $\frac{1}{2}$	**8**	$\frac{1}{2}$ of $\frac{1}{2}$ of $\frac{1}{2}$	**9**	$\frac{1}{2}$ of $\frac{1}{3}$ of $\frac{1}{4}$

10 If $p = \frac{1}{2}, q = \frac{1}{3}, r = \frac{1}{4}$ evaluate:

  a $pq + pr$ **b** $(p+q)r$ **c** $\frac{1}{2}p - \frac{1}{3}q$ **d** $pq + qr + rp$

11 At an election $\frac{4}{5}$ of the electorate voted. Of those who voted, $\frac{1}{2}$ supported the Labour candidate, $\frac{1}{3}$ the Conservative candidate and the remainder the Liberal candidate.

  a What fraction of those who voted supported the Liberal candidate?

  b What fraction of the total electorate voted for each of the three candidates?

  c Check that your three answers in *b* total $\frac{4}{5}$.

Simplify the following, where *a*, *b*, *c*, *p*, *q*, *x*, *y* represent natural numbers:

12a $\dfrac{2}{3}\times\dfrac{2}{5}$ *b* $\dfrac{2}{3}\times\dfrac{x}{5}$ *c* $\dfrac{3}{4}\times\dfrac{1}{x}$ *d* $\dfrac{p}{2}\times\dfrac{q}{3}$

  e $\dfrac{a}{4}\times\dfrac{3}{a}$ *f* $\dfrac{y}{4}\times\dfrac{6}{y}$ *g* $\dfrac{a}{b}\times\dfrac{b}{a}$ *h* $\dfrac{a}{b}\times\dfrac{b}{c}\times\dfrac{c}{a}$

Simplify the following:

13 $1\frac{1}{3}\times1\frac{1}{8}\times1\frac{1}{7}$ **14** $2\frac{5}{8}\times2\frac{1}{5}\times\frac{4}{7}$ **15** $\frac{3}{4}\times\frac{5}{6}\times\frac{8}{9}$

Use the commutative, associative, and distributive laws to help you to simplify:

16 $1\frac{1}{3}\times7\times\frac{3}{4}$ **17** $2\frac{1}{2}\times3\frac{3}{5}\times1\frac{1}{4}$

18 $(5\frac{1}{2}\times2\frac{3}{4})+(4\frac{1}{2}\times2\frac{3}{4})$ **19** $(3\frac{3}{8}\times2\frac{1}{3})+(3\frac{3}{8}\times1\frac{1}{3})$

20 $(1\frac{5}{6}\times3\frac{3}{4})-(1\frac{5}{6}\times3\frac{1}{4})$ **21** $(5\frac{7}{8}\times2\frac{3}{5})-(5\frac{7}{8}\times1\frac{3}{5})$

Division

Just as $6\div2$ asks the question 'How many *twos* are there in *six*?' so $6\div\frac{1}{2}$ asks the question 'How many *halves* are there in *six*?' That is, how many *halves* are there in *twelve halves*?

Thus $6\div\frac{1}{2} = \frac{12}{2}\div\frac{1}{2} = 12\div1 = 12.$

Example 1.

     a $2\frac{1}{4}\div\frac{3}{4} = \frac{9}{4}\div\frac{3}{4} = 9\div3 = 3$

     b Now $\quad 2\frac{1}{4}\times\frac{4}{3} = \dfrac{\overset{3}{\cancel{9}}}{\underset{1}{\cancel{4}}}\times\dfrac{\overset{1}{\cancel{4}}}{\underset{1}{\cancel{3}}} = 3$

Hence $\quad 2\frac{1}{4}\div\frac{3}{4} = 2\frac{1}{4}\times\frac{4}{3}$

Example 2.

     a $\frac{3}{4}\div\frac{5}{8} = \frac{6}{8}\div\frac{5}{8} = 6\div5 = \frac{6}{5} = 1\frac{1}{5}$

$$b \quad \text{Now} \quad \frac{3}{\cancel{4}} \times \frac{\cancel{8}^{2}}{5} = \frac{6}{5} = 1\tfrac{1}{5}$$

Hence $\tfrac{3}{4} \div \tfrac{5}{8} = \tfrac{3}{4} \times \tfrac{8}{5}$.

These examples illustrate the general rule that:

Division by $\dfrac{a}{b}$ is the same as multiplication by $\dfrac{b}{a}$.

Example 3. $\quad \dfrac{3}{10} \div 1\tfrac{1}{4} = \dfrac{3}{10} \div \dfrac{5}{4} = \dfrac{3}{\cancel{10}} \times \dfrac{\cancel{4}^{2}}{5} = \dfrac{6}{25}.$

Example 4. $\quad \dfrac{2\tfrac{3}{4}}{1\tfrac{1}{2}} = \dfrac{\tfrac{11}{4}}{\tfrac{3}{2}} = \dfrac{11}{\cancel{4}} \times \dfrac{\cancel{2}^{1}}{3} = \dfrac{11}{6} = 1\tfrac{5}{6}$

Exercise 7A

As in worked examples 1 and 2, calculate the values of *a* and *b* in each of the following, and check that they are equal:

1 *a* $\tfrac{3}{4} \div \tfrac{1}{2}$ *b* $\tfrac{3}{4} \times \tfrac{2}{1}$ 2 *a* $\tfrac{5}{6} \div \tfrac{1}{3}$ *b* $\tfrac{5}{6} \times \tfrac{3}{1}$

3 *a* $\tfrac{7}{8} \div \tfrac{3}{8}$ *b* $\tfrac{7}{8} \times \tfrac{8}{3}$ 4 *a* $\tfrac{2}{5} \div \tfrac{9}{10}$ *b* $\tfrac{2}{5} \times \tfrac{10}{9}$

5 *a* $\tfrac{2}{3} \div \tfrac{1}{2}$ *b* $\tfrac{2}{3} \times \tfrac{2}{1}$ 6 *a* $1\tfrac{1}{2} \div \tfrac{3}{4}$ *b* $1\tfrac{1}{2} \times \tfrac{4}{3}$

7 *a* $1\tfrac{1}{4} \div 1\tfrac{1}{2}$ *b* $1\tfrac{1}{4} \times \tfrac{2}{3}$ 8 *a* $\tfrac{3}{10} \div 2\tfrac{2}{5}$ *b* $\tfrac{3}{10} \times \tfrac{5}{12}$

Use the rule, as in worked examples 3 and 4, to do questions *9–20.*

9 $\tfrac{7}{10} \div \tfrac{3}{5}$ 10 $\tfrac{5}{8} \div \tfrac{3}{4}$ 11 $2\tfrac{1}{2} \div \tfrac{5}{8}$ 12 $2\tfrac{1}{3} \div 3\tfrac{1}{2}$

13 $3\tfrac{3}{8} \div 1\tfrac{1}{2}$ 14 $3\tfrac{1}{3} \div 2$ 15 $4\tfrac{2}{3} \div 3$ 16 $5\tfrac{1}{3} \div 1\tfrac{1}{3}$

17 $2 \div 1\tfrac{1}{2}$ 18 $4 \div \tfrac{3}{4}$ 19 $1 \div 2\tfrac{1}{2}$ 20 $3\tfrac{3}{4} \div \tfrac{3}{8}$

What fraction of

21 30 is 5?   22 6 is $1\tfrac{1}{2}$?   23 3 is $1\tfrac{1}{4}$?   24 $2\tfrac{5}{8}$ is $1\tfrac{3}{4}$?

Re-write each of the following, replacing ∘ by $+$, $-$, $\times$, or $\div$, so that the statement will be correct: (There may be more than one possible replacement)

25 $\tfrac{1}{2} \circ \tfrac{1}{2} = 1$ 26 $\tfrac{1}{2} \circ \tfrac{1}{2} = 0$ 27 $\tfrac{1}{2} \circ \tfrac{1}{2} = \tfrac{1}{4}$ 28 $1 \circ \tfrac{1}{4} = 1\tfrac{1}{4}$

29 $1 \circ \frac{1}{4} = 4$ 30 $1 \circ \frac{1}{4} = \frac{3}{4}$ 31 $1 \circ \frac{1}{4} = \frac{1}{4}$ 32 $\frac{1}{2} \circ \frac{1}{3} = \frac{1}{6}$

33 $\frac{1}{2} \circ \frac{1}{3} = 1\frac{1}{2}$ 34 $\frac{1}{2} \circ \frac{1}{3} = \frac{5}{6}$ 35 $\frac{4}{3} \circ \frac{3}{4} = 1$ 36 $\frac{4}{3} \circ \frac{3}{4} = \frac{16}{9}$

37 $\frac{4}{3} \circ \frac{3}{4} = \frac{7}{12}$ 38 $\frac{4}{3} \circ \frac{3}{4} = \frac{25}{12}$

Exercise 7B

Simplify each of the following:

1 $\frac{7}{8} \div \frac{1}{2}$ 2 $1\frac{2}{3} \div \frac{5}{6}$ 3 $4\frac{2}{3} \div 1\frac{7}{9}$ 4 $4\frac{1}{5} \div \frac{7}{8}$

5 $2\frac{4}{7} \div 1\frac{13}{14}$ 6 $5\frac{1}{4} \div 2$ 7 $3 \div 2\frac{1}{3}$ 8 $1 \div 1\frac{3}{8}$

9 $(1\frac{1}{4} + 2\frac{1}{8}) \div 2\frac{1}{4}$ 10 $(3\frac{1}{2} + 4\frac{5}{8}) \div 3\frac{1}{3}$ 11 $(2\frac{1}{3} - 1\frac{1}{2}) \div 1\frac{1}{3}$

What fraction of

12 $2\frac{1}{2}$ is $\frac{1}{2}$?   13 $6\frac{1}{4}$ is $1\frac{1}{4}$?   14 $3\frac{3}{4}$ is $3\frac{1}{8}$?   15 $1\frac{1}{3}$ is $\frac{5}{6}$?

Rewrite each of the following, replacing ∘ by $+$, $-$, $\times$, or $\div$ so that the statement will be correct:

16 $\frac{3}{4} \circ \frac{1}{2} = \frac{1}{4}$ 17 $\frac{3}{4} \circ \frac{1}{2} = \frac{3}{8}$ 18 $\frac{3}{4} \circ \frac{1}{2} = 1\frac{1}{4}$

19 $\frac{3}{4} \circ \frac{1}{2} = 1\frac{1}{2}$ 20 $1\frac{1}{2} \circ 2\frac{1}{3} = 3\frac{5}{6}$ 21 $\frac{7}{8} \circ 1\frac{1}{3} = 1\frac{1}{6}$

22 $(4 - 2\frac{1}{4}) \circ 1\frac{1}{3} = 2\frac{1}{3}$ 23 $x \circ x = 1$ 24 $x \circ x = 0$

Use the symbols $+$, $-$, $\times$, $\div$, $=$, to make at least 3 true statements about each of the following sets of 3 fractions:

25 $\frac{1}{2}, \frac{1}{4}, \frac{3}{4}$ 26 $\frac{1}{2}, \frac{2}{3}, \frac{3}{4}$.

If $a = 8$, $b = 1\frac{1}{2}$, $c = \frac{3}{4}$ find the values of:

27 ab 28 $a \div b$ 29 $b \div a$ 30 $b \div c$

31 $(ab) \div c$ 32 $(ac) \div b$ 33 $1 \div (abc)$ 34 $(2b) \div (3c)$

35 Divide the sum of $\frac{1}{2}$ and $\frac{1}{3}$ by their difference.

5 Ratios

If the number of pupils on the roll of a class is 39 and 3 are absent, then the ratio of the number absent to the number present is $\frac{3}{36}$ or $\frac{1}{12}$.

What is the ratio of

a the number absent to the total roll?
b the number present to the total roll?

The ratio of a to b, where b is not zero, is $a \div b$, or $\dfrac{a}{b}$, and is often written $a : b$.

Note.—We saw in Section 3 (page 193) that every number in the form of a fraction $\frac{a}{b}$, and hence in the form of a *ratio* $a:b$, can be placed on the number line; all such numbers are called *rational numbers*.

In finding the ratio of two quantities we must express them in the same units. What is the ratio of 1 m to 1 km? Of 1 penny to £1? Of 1 minute to 1 hour?

Example.—Express these ratios in their simplest forms:

a $\frac{3}{4}:2$ *b* $1\frac{1}{4}:1\frac{2}{3}$ *c* $87\frac{1}{2}$p : £1·50

a $\frac{3}{4}:2 = \frac{\frac{3}{4}}{2} = \frac{3}{4}\times\frac{1}{2} = \frac{3}{8} = 3:8$

b $1\frac{1}{4}:1\frac{2}{3} = \frac{\frac{5}{4}}{\frac{5}{3}} = \frac{5}{4}\times\frac{3}{5} = \frac{3}{4} = 3:4$

c $87\frac{1}{2}$p : £1·50 $= \frac{87\frac{1}{2}}{150} = \frac{175}{2\times150} = \frac{7}{2\times6} = \frac{7}{12} = 7:12$

Exercise 8A

Express each of the following ratios in its simplest form:

1   350 : 400   *2*   49 : 245   *3* $1\frac{1}{2}:3$

4 $2\frac{1}{2}:3\frac{3}{4}$ *5* $12\frac{1}{2}$ pence : £1   *6*   25p : £5

7   80 cm : 1 m   *8*   8 mm : 6 mm   *9*   45 minutes : $2\frac{1}{4}$ hours

10   The ratio of the top number to the number below it in the table is the same in every case. Fill in the blanks.

4	8	20			1	$2\frac{1}{2}$
6			36	60		

11   Two rectangles have lengths 8 m and 4 m, and breadths 6 m and 3 m respectively. Find the ratios of:

a their lengths   *b* their breadths   *c* their areas.

12   Two squares have sides of lengths 20 cm and 15 cm. Find the ratios of:

a the lengths of the sides   *b* the areas of the squares.

13   Two squares have sides of lengths 8 m and 4 m. What is the ratio of their areas? Can you find this answer without working out each area?

14 In Figure 12, each large shape is divided into equal parts.
Give the ratio of the shaded part to the whole shape in each case.

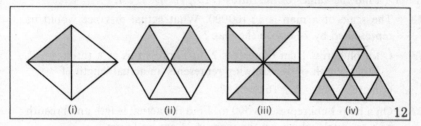

| (i) | (ii) | (iii) | (iv) | 12 |

Find the ratios of the following prices (noticing that you must compare the same quantity in each pair):

15 12½ pence each and £1·25 per dozen.

16 £1·50 per dozen and 17½ pence each.

17 2½p per metre and £40 per km.

Exercise 8B

Express each of the following ratios in its simplest form:

1 56 : 144   *2* 1⅔ : 4   *3* 3⅓ : 2⅞   *4* 47½p : £1

5 8p each : £1·44 per dozen   *6* 2p for 3 : 3p for 2

7 The following table compares two cars under 5 different headings: Find the ratio 'car A : car B' in each case.

	Cost	Length	Engine Capacity	Fuel Tank	Petrol Consumption
Car A	£575	350 cm	875 cm^3	27 litres	16 km per litre
Car B	£1000	420 cm	1300 cm^3	45 litres	12 km per litre

8 If the two cars in question *7* undertake the same journey of 180 km non-stop, and if car A averages 48 kph and car B 72 kph, find the ratio of their times. Compare this with the ratio of their speeds.

9 The population of a town is 35 280 of whom 18 900 are female. What is the ratio of males to females?

10 In an examination, 15 pupils out of 35 scored less than half-marks. What is the ratio of the number of pupils who scored half-marks or more to the number who scored less than half-marks?

11   On a certain Saturday the number of football matches that ended in home wins (H) was 25, away wins (A) 15, draws (D) 10.
     Find the values of the ratios $H:A$, $H:D$, $D:A$.

12   The scale of a map is $1:100000$. What actual distance would be represented by $2\frac{1}{2}$ cm on the map?

13   On a blueprint 1 cm represents 2 m. Express this as a ratio.
     What length on the drawing represents an actual length of

  a 6 metres   *b* 16 metres?

14   On a map 1 cm represents 500 m. Find the actual length and breadth of a rectangle which, on the map, is $2\frac{1}{2}$ cm by $1\frac{1}{2}$ cm.

15   The length and breadth of a room are in the ratio $5:4$. Find the length if the breadth is 3·6 metres.

16   Two cubes have edges 1 metre and 2 metres long. Find the ratios of:

  a the lengths of their edges   *b* the areas of their faces
  c their volumes.
     What do you notice?

17   Repeat question *16* for two cubes of edges 8 cm and 12 cm.

---

6 An introduction to percentages

Suppose that in a series of tests a boy obtained the following results:

	English	History	Geography	Mathematics	Science
Actual mark	15	76	37	60	28
Possible mark	20	100	50	75	40

In which subject does he obtain the best result? In which the worst?

Does the fact that in history he got the highest mark mean that it is his best result?

If we write down the ratios of his actual scores to the possible scores we obtain $\frac{15}{20}, \frac{76}{100}, \frac{37}{50}, \frac{60}{75}, \frac{28}{40}$.

These simplify to $\frac{3}{4}, \frac{19}{25}, \frac{37}{50}, \frac{4}{5}, \frac{7}{10}$, and it is difficult to see which of these is the largest fraction and which is the smallest.

Let us arrange that all the denominators are the same, say 100.

Then $\dfrac{3}{4} = \dfrac{3 \times 25}{4 \times 25} = \dfrac{75}{100}$. What are the others? Can you now see which are his best and worst subjects?

It is often very useful to express fractions so that their denominators are 100.

A fraction with denominator 100 is called a *percentage*. Per cent means per hundred.

We write $\frac{75}{100} = 75$ per cent $= 75\%$.

Also $10\% = \frac{10}{100} = \frac{1}{10}$.

Example 1. $62\frac{1}{2}\% = \dfrac{62\frac{1}{2}}{100} = \dfrac{125}{200} = \dfrac{5}{8}$

Example 2. 115% of £3 $= \dfrac{115}{100} \times £3 = £\dfrac{345}{100} = £3 \cdot 45$

Exercise 9

Express the following percentages as fractions in their simplest form:

1	20%	*2*	30%	*3*	40%	*4*	15%	*5*	100%
6	50%	*7*	175%	*8*	48%	*9*	16%	*10*	24%
11	2%	*12*	12½%	*13*	45%	*14*	105%	*15*	95%
16	2½%	*17*	33⅓%	*18*	150%	*19*	66⅔%	*20*	87½%

Find the values of:

21 12% of £50

22 35% of 150, and of £500

23 5½% of 800

24 2½% of £120, and of 50 kg

25 120% of 10 metres

26 6⅔% of 3 km, and of £60

27 2½% of 20

28 160% of 12 metres, and of £2

29 12½% of 4 litres

30 16⅔% of 12 hours, and of £33

31 A discount of 15% was given during a sale. Find the discount on goods priced at £15·60.

32 If a school roll of 860 pupils rose by 5%, what was the new roll?

33 A salesman was allowed 7½% commission on his sales. Find his commission on sales totalling £312.

     *   *   *   *

Some percentages occur so often that it is worth while memorizing their simplest values as fractions, and vice versa, e.g.

$$2\tfrac{1}{2}\% = \tfrac{1}{40}$$
$$5\% = \tfrac{1}{20} \qquad 25\% = \tfrac{1}{4}$$
$$10\% = \tfrac{1}{10} \qquad 50\% = \tfrac{1}{2} \qquad 33\tfrac{1}{3}\% = \tfrac{1}{3} \qquad 100\% = 1$$
$$20\% = \tfrac{1}{5} \qquad 75\% = \tfrac{3}{4} \qquad 66\tfrac{2}{3}\% = \tfrac{2}{3}$$

Note also: 5% of £1 = 5 pence and $2\tfrac{1}{2}\%$ of £1 = $2\tfrac{1}{2}$ pence

Using the above information, find the values of:

34	$2\tfrac{1}{2}\%$ of £200	**35**	50% of 45 cm	**36**	$33\tfrac{1}{3}\%$ of £6·30
37	25% of 500 ml	**38**	$66\tfrac{2}{3}\%$ of 12 km	**39**	100% of £13·57
40	10% of 700 m²	**41**	75% of 40 kg	**42**	20% of 800 g
43	3% of £1	**44**	$6\tfrac{1}{2}\%$ of £1	**45**	123% of £1

To express a fraction as a percentage

Make the denominator of the fraction 100.

Example
$$\frac{3}{5} = \frac{3 \times 20}{5 \times 20} = \frac{60}{100} = 60\%$$

or
$$\frac{3}{5} = \frac{\tfrac{3}{5} \times 100}{100} = \frac{60}{100} = 60\%$$

This can be written shortly

$$\tfrac{3}{5} = (\tfrac{3}{5} \times 100)\% = 60\%.$$

To change a fraction to a percentage, multiply it by 100.

Exercise 10

Express each of the following fractions as a percentage:

1	$\tfrac{1}{2}$	*2*	$\tfrac{1}{4}$	*3*	$\tfrac{2}{5}$	*4*	$\tfrac{3}{4}$
5	$\tfrac{4}{5}$	*6*	$\tfrac{3}{50}$	*7*	$\tfrac{19}{25}$	*8*	$\tfrac{1}{8}$
9	$\tfrac{10}{10}$	*10*	$\tfrac{5}{2}$	*11*	$\tfrac{1}{100}$	*12*	$\tfrac{1}{12}$

13 A class contained 15 girls and 20 boys. What percentage of the class are *a* girls *b* boys?

Can you find a quick way to work out *b*?

14   A factory had 1200 workers. 80 were absent. What percentage were on duty?

15   A girl leaves home at 8 am and returns at 4 pm. What percentage of 24 hours is she away from home?

16   A rugby team played 36 games and won 30 of them. What percentage did it win?

17   A boy got 60 marks out of 75 in a test. What percentage was this?

18   A car was bought for £500 and sold for £375. Find what percentage the loss was of the cost price.

19   For all her first-year examinations a girl got 385 marks out of a possible 550. What percentage was this?

20   There are 24 boys in a class of 36 pupils. What percentage of the class are girls?

   On a certain day 3 boys and 2 girls are absent. Find the percentage absence for   *a* boys   *b* girls   *c* the whole class.

Exercise 11 Miscellaneous questions on percentages

1   A man earning £24 per week was given an increase of 5%. What is his new weekly wage?

2   £720 was divided between two persons so that one was given 45% of the total. How much did each receive?

3   Shoes are marked at £2·40. Find the actual price paid if a discount of $2\frac{1}{2}\%$ is allowed.

4   A man has a salary of £1800. He spends $12\frac{1}{2}\%$ of it on food, 25% on his house, 8% on holidays, 10% on insurance. He pays 24% in income tax, and saves the rest.

   Find the actual amount of money he saves, spends, and pays on the various items given.

5   A firm exported machinery to the value of £750000 in 1968. The following year it increased this figure by 4%; what was the value then?

   In 1970 the firm planned to increase the 1969 figure by 5%; what would the value of its exports be in 1970?

6   An aircraft flew 5% faster than the speed of sound (1200 kilometres per hour at sea level). What was its speed?

7   A man bought a new car for £1200 and after a year it was worth
     only £1050.
     Express the depreciation (fall in value) as a percentage of
  a the price when new   b   the second-hand price.

8   A rectangle measures 15 cm by 5 cm. Find its area.
         If the length and the breadth are both increased by 20%, express
     the new area as a percentage of the original area. What is the
     percentage increase?

9   A man earns £2000 a year and saves 6% of this. Out of the remainder
     he pays £230 in income tax. Express what is left as a percentage of
     his £2000 salary.

10   Find out what percentage of the pupils in your class:
  a play football, rugby, or hockey
  b like 'pop' music
  c watch certain TV programmes (decide which ones)
  d enjoy arithmetic.

Illustrative summary

1 *The meaning of a fraction*

$$\frac{a}{b} = a \div b$$

  a is the *numerator* of the fraction
  b is the *denominator* of the fraction

2 *Equal fractions*

$$\frac{1}{2} = \frac{1 \times 3}{2 \times 3} = \frac{3}{6} = \frac{3 \times 4}{6 \times 4} = \frac{12}{24} = \frac{12 \div 2}{24 \div 2} = \frac{6}{12} = \cdots$$

3 *Least common multiples*

  The LCM of *a* 2, 3 is 6   *b* 2, 3, 4 is 12   *c* 5, 10 is 10

4 *Addition and subtraction of fractions*

$$\tfrac{3}{4} + \tfrac{5}{6} = \tfrac{9}{12} + \tfrac{10}{12} = \tfrac{19}{12} = 1\tfrac{7}{12}$$

5 *Multiplication of fractions*

$$\frac{3}{4} \text{ of } \frac{5}{6} = \frac{\overset{1}{\cancel{3}}}{4} \times \frac{5}{\underset{2}{\cancel{6}}} = \frac{5}{8}$$

6 *Division of fractions*

$$3\tfrac{1}{3} \div 2\tfrac{1}{2} = \frac{10}{3} \div \frac{5}{2} = \frac{10}{3} \times \frac{\overset{2}{2}}{\underset{1}{\cancel{5}}} = \tfrac{4}{3} = 1\tfrac{1}{3}$$

$$\frac{p}{q} \div \frac{a}{b} = \frac{p}{q} \times \frac{b}{a}$$

7 *Ratios*

$$a : b = \frac{a}{b}$$

8 *Percentages*

$$1 \text{ per cent} = 1\% = \frac{1}{100}$$

$$a\% = \frac{a}{100}$$

$$\frac{x}{y} = \frac{x}{y} \times 100\%$$

Topic to explore. Even and odd numbers

1   Copy and complete this addition table. Remember to take the first number from the left side of the table, the second number from top, and the answer where the row and column meet.

+	2	3
2		
3		

    From the answers in your table, do you obtain even or odd numbers when you add

a two even numbers

b two odd numbers

c an even and an odd number?

2   Let E represent an even number and let O represent an odd number. Copy and complete this table, filling Es and Os in the blank squares.

+	E	O
E		
O		

    Using the letters E and O fill in the answers to:

a $E+O =$

b $O+E =$

c Is $E+O = O+E$?

d What law of addition does this suggest?

e $(E+O)+E = \quad + \quad =$

f $E+(O+E) = \quad + \quad =$

g Is $(E+O)+E = E+(O+E)$?

h What law of addition does this suggest?

3   Copy and complete this multiplication table, where E represents an even number and O represents an odd number.

×	E	O
E		
O		

Fill in the answers to the following:

a $E \times O =$

b $O \times E =$

c Is $E \times O = O \times E$?

d What law of multiplication does this suggest?

e $(E \times O) \times E = \quad \times \quad =$

f $E \times (O \times E) = \quad \times \quad =$

g Is $(E \times O) \times E = E \times (O \times E)$?

h What law of multiplication does this suggest?

i $(E \times O)+(E \times E) = \quad + \quad =$

j $E \times (O+E) \quad = \quad \times \quad =$

    k Is $(E \times O) + (E \times E) = E \times (O + E)$?

    l What law of multiplication and addition does this suggest?

4   The table represents the results of a certain operation * on some pairs of numbers from the set {0, 1, 2, 3, 4}.

    a Copy the table and fill in the missing numbers when you have recognized the operation *.

    b List some interesting facts about the table.

    c Is the operation * commutative?

*	0	1	2	3	4
0	.	0	0	0	.
1	.	.	4	6	8
2	0	4	.	12	.
3	0	6	.	.	24
4	0	8	16	.	.

Revision Exercises

Revision Exercises on Chapter 1
The System of Whole Numbers

Revision Exercise 1A

1 Write down the meaning of each of the following numbers, then work out its value:

a 0^2, 1^2, 8^2, 11^2, 15^2, 30^2, 50^2, 100^2.
b 0^3, 1^3, 3^3, 10^3.

2 List the following sets:

a The set of whole numbers less than 12.
b The set of even numbers between 49 and 63.
c The set of square numbers less than 90.
d The set of even prime numbers.

3 Write down the next two members of the sequences which start with the numbers given:

a 1, 8, 15, 22, ...   *b* 2, 3, 5, 8, 12, ...
c 3, 4, 6, 10, 18 ...   *d* 100, 99, 98, 97, ...
e 1, 11, 21, 31, ...   *f* 64, 32, 16, 8, ...

4 Write down suitable missing numbers in these sequences:

a 12, 10, 8, ..., 4, 2, ...   *b* 1, 2, 4, 7, 11, ..., ...
c 4, 9, ..., 25, 36   *d* 0, 1, 8, ..., 64, ...

5 Complete these magic squares:

11	6	13
		8

a

4	9		
5		3	
	2		
14	7	12	1

b

8		4
	5	9
		2

c

6   If * means 'add the first number to twice the second number' find the values of:

   a   3 * 4 and 4 * 3   *b*   5 * 0 and 0 * 5

   c   (1 * 2) * 3 and 1 * (2 * 3)   *d*   (3 * 0) * 5 and 3 * (0 * 5)

        Is the operation * commutative?

        Is the operation * associative?

7   Find the LCM of:

   a   2, 7   *b*   2, 6   *c*   6, 10   *d*   2, 5, 6   *e*   4, 5, 6

8   Write down all the prime numbers between 30 and 40.

9   Express as products of prime factors:

   a   18   *b*   30   *c*   84   *d*   75

10   Evaluate in the easiest way you can:

   a   75 + (49 + 25)   *b*   (11 + 72) + 39   *c*   2 × 99 × 5

   d   (25 × 53) × 4   *e*   (17 × 5) + (3 × 5)   *f*   (16 × 9) + (11 × 16)

Revision Exercise 1B

1   Evaluate   *a* $10^2, 25^2, 100^2, 1000^2$

                 b $0^3, 1^3, 5^3, 10^3, 100^3, 1000^3$

2   List the following sets:

   a   The set of whole numbers greater than 100 and less than 110.

   b   The set of odd numbers between 90 and 100.

   c   The set of square numbers which are less than 200.

   d   The set of prime numbers between 40 and 50.

3   Add two terms to continue each of the following sequences:

   a   3, 8, 13, 18, ...   *b*   35, 28, 21, 14, ...

   c   0, 1, 8, 27, ...   *d*   3, 4, 7, 12, 19, ...

   e   1, 9, 2, 8, 3, 7, 4, 6, ...   *f*   0, 2, 6, 12, 20, ...

4   If △ means 'square the first number and add the second' find the values of:

   a   10 △ 1 and 1 △ 10   *b*   3 △ 0 and 0 △ 3

5   If ⊕ means 'increase the first number by 2 and then multiply by the second number', evaluate:

   a   3 ⊕ 5   *b*   5 ⊕ 3   *c*   4 ⊕ 7   *d*   7 ⊕ 4

   e   (6 ⊕ 2) ⊕ 3   *f*   (0 ⊕ 5) ⊕ 4   *g*   (5 ⊕ 0) ⊕ 4

        Is the operation ⊕ commutative?

Suppose ∘ means 'square the second number and multiply the answer by the first number'. Evaluate:

a 2 ∘ 3   *b* 3 ∘ 2   *c* 0 ∘ 7   *d* 7 ∘ 0
e 1 ∘ 4   *f* 4 ∘ 1   *g* (1 ∘ 2) ∘ 5   *h* 1 ∘ (2 ∘ 5)

Is the operation ∘ commutative?
Is the operation ∘ associative?

7   Find the LCM of:

a 15, 20   *b* 2, 3   *c* 2, 3, 4   *d* 2, 3, 4, 5   *e* 2, 3, 4, 5, 6

8   Find the smallest natural number which is divisible by 3 and also by the next two greater prime numbers.

9   Express as a product of prime factors: *a* 63   *b* 48   *c* 1260

10   Use the commutative and associative laws of addition and multiplication, and the distributive law of multiplication over addition to work out each of the following in the easiest way you can:

a $17 + 89 + 23$ *b* $25 \times 13 \times 4$ *c* $8 \times 17 \times 125$
d $(13 \times 6) + (6 \times 7)$ *e* $15 \times 23 \times 4$ *f* $(37 \times 43) + (57 \times 37)$

11   Find the answers to the first three of the following multiplications by actually multiplying; then write down the answers to the others:

37×3 37×6 37×9 37×12 37×15
37×18 37×21 37×24 37×27

Can you explain your method?

12   Work out the first three of the following, then write down the answers to the others:

$$1 \times 1 =$$
$$11 \times 11 =$$
$$111 \times 111 =$$
$$1111 \times 1111 =$$
$$11111 \times 11111 =$$
$$111111 \times 111111 =$$

13   Divide 2184 by 28 using long division. Now divide 2184 by 4 and divide the answer by 7. Which way is easier?

Work out the following the easiest way you can:

a $\dfrac{13\,716}{36}$ *b* $\dfrac{18\,590}{110}$ *c* $\dfrac{16\,002}{63}$

14*a*   Multiply 37 by 5. Now multiply 37 by 10 and divide the answer by 2. Why do you get the same result?

b Calculate the value of 5490×5, first by multiplying by 5, second by multiplying by 10 and dividing the answer by 2. Which method is easier?

c Evaluate 7849×5 and 58736×5

15 Can you think of an easy way to multiply by 25? Multiply these by 25:

a 4896   b 7862   c 3791.

16 In which of the following may the order of events be commuted (i.e. the first and second event interchanged) without necessarily affecting the result?

a Answer question 4 in the examination paper, and then answer question 5.

b In a motor-boat sail 4 miles north, and then 3 miles east.

c Put the car in the garage, and then lock the door.

d Buy a suit, and then buy a hat.

e Multiply by 3, and then multiply by 13.

f Interchange the units and tens digits, and then double the number.

17 A number is said to be 'perfect' when it is equal to the sum of all its divisors, excluding the number itself. For example, $6 = 1 + 2 + 3$.
   There is another 'perfect' number less than 32; find it.

18 Which of the following are factors of $836\,748\,231$?

a 2   b 3   c 5   d 9   e 11.

19 Look at this pattern: $(2 \times 2) - (1 \times 3) = 1$
   $$(3 \times 3) - (2 \times 4) = 1$$
   $$(4 \times 4) - (3 \times 5) = 1$$

Use these lines to complete the statement $(193 \times 193) - \ldots = 1$, and check your answer. Use the pattern to write down the value of 999×999.

20 If x, y, and z represent whole numbers, state whether each of the following is true or false:

a $x \times y = y \times x$ b $x \div y = y \div x$
c $(x \times y) \times z = (z \times y) \times x$ d $x - y = y - x$
e $x + (y \times z) = (x + y) \times z$ f $(x \times y) \div z = x \times (y \div z)$

Revision Exercises on Chapter 2
Decimal Systems of Money, Length, Area, Volume, and Mass

Revision Exercise 2A

1   Find the cost of:
a   24 daily newspapers at $2\frac{1}{2}$p each.
b   5 records at $42\frac{1}{2}$p each.
c $\frac{3}{4}$ kg of beef at £1·20 per kg.
d   3 boxes of chocolates at 67p each.
e   28 bus tickets at 16p each.
f   All the items in *a* to *e* added together.

2   How much change should be received from a £5 note after payment of 3 articles at 45p each, 2 at 75p each, and 6 at $33\frac{1}{2}$p each?

3   Calculate the perimeters of rectangles with lengths and breadths:
a   15 mm and 5 mm   b   16 cm and 7 cm   c   550 m and 375 m

4   I take 132 steps, each 80 cm long, in pacing the length of a football pitch. What is the length of the pitch in metres?

5   Telephone poles at the side of a straight road are 65 m apart. What is the distance between the first and sixteenth poles?

6   17 volumes of an encyclopaedia, each 5 cm thick, and a dictionary 7 cm thick, are placed on a shelf 1 metre long.
How many books, each 2 cm thick, will fill the space that is left?

7   The distance round a running track is 200 m. How many laps of the track will there be in the 5000 m race?

8   2500 sheets of card, each $\frac{1}{2}$ mm thick, are put in a pile. What is the height of the pile:
a   in mm   b   in cm   c   in m?

9   Find the perimeters of the rectangles with lengths and breadths:
a   2 cm and 6 mm   b   5·50 m   and 3·50 m

10 Find the perimeters of squares with lengths of side:
a   35 cm   b   150 cm

11 How many edging slabs, each 80 cm long, are required to edge a square lawn of side 9 m 60 cm?

12 A playing field is 400 m long and 300 m broad. What is the total cost of enclosing the field with a fence costing £1·50 per metre?

13 By tracing Figure 1, and placing your tracing over 5-mm squared paper, estimate the number of squares it encloses.

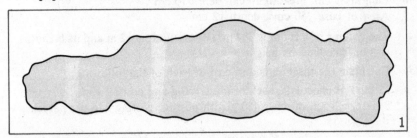

14 Calculate the areas of these rectangles:

	a	b	c	d	e
Length	20 cm	18 mm	20 m	85 mm	13 km
Breadth	16 cm	18 mm	17 m	6 mm	9 km

15 A road is 800 m long and 12 m wide. What is the cost of surfacing it at £1·20 per m²?

16 A bowling green is a square of side 35 m. What is the cost of returfing it at 60p per m²?

17 Calculate the shaded areas in Figures 2 and 3.

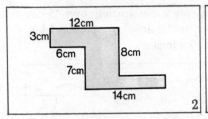

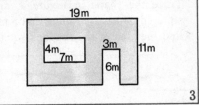

18a How many metres are there in 1 kilometre? So how many square metres are there in 1 square kilometre?

  b If the area of a hockey pitch is 4000 m², what fraction of a square kilometre does it cover?

  c A square field has a side 500 m long. What fraction of a square kilometre does it cover?

19 Find the volumes of cuboids with lengths, breadths and heights:

  a 6 cm, 5 cm, and 3 cm   b 4 m, 2 m, and 50 cm

20 Find the volume of cubes with lengths of side:

   a 10 cm   *b*   8 m   *c* 5 mm   *d* 25 m

21 Find the volume of water in litres in rectangular tanks as follows:

   a Length 25 cm, breadth 20 cm, depth 15 cm

   b Area of base 750 cm², depth 12 cm

22 The volume of a room is 384 m³. If its length is 12 m and its breadth is 8 m, calculate its height.

23 Calculate the total mass and cost of each of the following:

   a 12 bars of chocolate, each of mass 150 g and costing 13p.

   b 8 parcels, each of mass 450 g with postage costing 24p per parcel.

Revision Exercise 2B

1 Running costs for a year for a motor car were as follows:
Petrol £60·50; oil £1·25; repairs £18·82; garaging £12; licence £25.
Calculate the total cost, and the cost per kilometre to the nearest halfpenny if the car ran 8000 km during the year.

2 Sound travels at 330 m per second. If thunder is heard 24 seconds after the lightning is seen, how far away is the storm? What had you to assume in finding the answer?

3 What length of wire would be needed to construct a skeleton cuboid 25 cm long, 18 cm broad, and 7·5 cm high?

4 Trace the shape in Figure 4, and place the tracing over 5-mm squared paper. Assuming that 5 mm on the drawing represent 1 km, find the area of the shape in square kilometres.

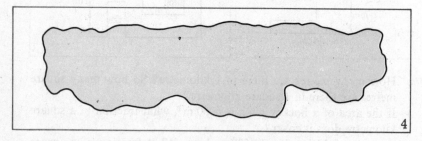

4

5 *a* What is the area of a rectangular lawn 44 m long and 35 m broad?

   b How many kg of grass seed are necessary to sow the lawn, at the rate of 50 g per m²?

   c What is the cost of the grass seed at 90p per kg?

6   The rectangular floor of a room is 6 m long and 4 m broad. A carpet 5 m by 4 m covers part of it, and the rest is covered with linoleum. Find the total cost if the carpet costs £3·50 per m², and the linoleum 45p per m².

7   240 square tiles, each of side 25 cm, are needed to cover a rectangular floor. If the floor is 5 m long, what is its breadth?

8   How many paving stones, each 75 cm by 60 cm, are needed to cover a rectangular courtyard 9 m by 6 m?

9   By making sketches, find the largest number of rectangular tiles, all lying the same way, which can be fitted into the given rectangles:

  a   Tiles 6 cm by 4 cm; rectangle 36 cm by 32 cm.
  b   Tiles 8 cm by 5 cm; rectangle 48 cm by 32 cm.
  c   Tiles 9 cm by 4 cm; rectangle 31 cm by 21 cm.

10   Four main streets in an American city enclose a rectangular block 800 m by 500 m. Express the area of the block as a fraction of 1 km².

11   The area of a housing scheme is to be 1 km². If each house is allowed 500 m², how many houses can be built?

12   A rectangular water tank *without a lid* is to be made from sheet metal, and is to be 5 m long, 3 m wide, and 2 m deep.

  a   What area of sheet metal is required?
  b   What volume of water will the tank hold?

13   A wagon 4 m by 2 m by 1½ m weighs 3000 kg. It is filled level with coal of mass 1500 kg per m³. Calculate the total mass of wagon and coal.

14   Bricks 5 cm by 3 cm by 2 cm are to be packed into a rectangular box 23 cm by 17 cm by 14 cm, so that all bricks lie the same way.
    Calculate the number of bricks that can be put into the box, in each of the six ways possible.

15   A rectangular water tank weighs 30 kg when empty. The area of the base of the tank is ½ m², and water enters to a depth of 1 m. What is the weight of tank plus water, given that the mass of water is 1000 kg per m³?

16   The surface of a reservoir has an area of 4000 m², and the depth of water is 2 m. Given that the mass of water is 1000 kg per m³, and that 1 tonne = 1000 kg, find the mass of water in the reservoir in tonnes.

Revision Exercises on Chapter 3
Fractions, Ratios and Percentages

Revision Exercise 3A

1. Simplify: *a* $\frac{3}{6}$ *b* $\frac{10}{12}$ *c* $\frac{16}{20}$ *d* $\frac{16}{36}$ *e* $\frac{15}{15}$

2. Find the values of:
   a $\frac{2}{3}$ of £474 *b* $\frac{2}{5}$ of 1 hour *c* $\frac{7}{20}$ of 1 litre

3. By noticing that $\frac{4}{5} = 1 - \frac{1}{5}$, $\frac{7}{8} = 1 - \frac{1}{8}$, and so on, evaluate:
   a $\frac{4}{5}$ of £725 *b* $\frac{7}{8}$ of 2 hours *c* $\frac{3}{4}$ of £7·68

4. Find the total cost of the following:
   2 kg sugar at 19p per kg; $\frac{1}{4}$ kg butter at 38p per kg;
   250 g tea at 64p per kg; $1\frac{1}{2}$ dozen eggs at $2\frac{1}{2}$p each egg.

5. Give the LCM of *a* 3, 4 *b* 3, 6 *c* 2, 5 *d* 4, 6

6. The first 75 units of electricity I use are charged at $2\frac{1}{2}$ pence per unit, and the rest at $\frac{1}{2}$ penny per unit. What is my bill when I have used 1275 units in a quarter?

7. Simplify:
   a $\frac{3}{4} + \frac{1}{6}$ *b* $\frac{2}{3} - \frac{5}{8}$ *c* $1\frac{1}{2} + 3\frac{3}{4}$ *d* $4\frac{1}{3} - 2\frac{1}{5}$

8. Simplify:
   a $\frac{3}{4}$ of $\frac{2}{9}$ *b* $5\frac{1}{3} \times 1\frac{1}{4}$ *c* $\frac{7}{8} \div \frac{3}{4}$ *d* $2\frac{5}{8} \div 1\frac{1}{6}$

9. Find the volumes of the following cuboids:
   a Length 4 metres, breadth 3 metres, height $2\frac{1}{2}$ metres
   b Length 6 mm, breadth $4\frac{1}{2}$ mm, height $1\frac{1}{2}$ mm
   c Area of cross-section 3 cm², height $4\frac{1}{2}$ cm
   d Area of cross-section $1\frac{1}{2}$ m², length $17\frac{1}{2}$ m

10. It is required to make a model of a cube of edge $3\frac{1}{2}$ cm.
    a What length of wire is necessary to make a skeleton cube?
    b What area of cardboard is necessary to make a cardboard model?

11. If * means 'take half the first number and add one quarter of the second number', calculate the values of:
    a $4 * 6$ and $6 * 4$ *b* $\frac{1}{4} * 2$ and $2 * \frac{1}{4}$

12a. Which of the following fractions is nearest to $3\frac{1}{2}$?
    $$3\frac{3}{8}, \quad 3\frac{9}{16}, \quad 3\frac{5}{12}.$$

   b Find the difference between the greatest and the least of the 3 fractions.

13   Find in the simplest form the ratio of:

  a $37\frac{1}{2}$p to $87\frac{1}{2}$p   *b*   450 m to 1 km

14   One man is paid £20 per week; a second is paid £65 per month. Find the ratio of their annual incomes.

15   Two cubes have edges of length 2 cm and 3 cm. Find the ratio of:

  a   the total lengths of their edges   *b*   the total areas of their faces
  c   their volumes.

16   Express the following as fractions of £1, and as percentages of £1:

  a   30p   *b* $1\frac{1}{2}$p   *c*   95p   *d*   £0·75   *e*   £0·$87\frac{1}{2}$

17   Find the amount paid in tax (at the percentage rate stated in brackets) on the articles whose wholesale prices are given:

  a   £2·50 (10%)   *b*   £180 ($33\frac{1}{3}$%)   *c*   £0·72 ($12\frac{1}{2}$%)

18   The roll of a school is 1250. On a day when 8% of the pupils were absent, how many were present?

19   A suit is marked at £15·50. At a sale, 10% is deducted from the price. At what price is it sold?

20   A man who earns £24 per week gets a rise of $7\frac{1}{2}$%. What is his new wage?

21   In a school there are 224 boys and 288 girls. What percentage of the total roll is boys, and what percentage is girls?

22   In one class 3 pupils were absent out of 30 and in another class 4 out of 36. Find for each class:

  a   the percentage of pupils absent.
  b   the percentage of pupils present.

23   At an election there were two candidates and 37 500 voters on the roll. The result of the poll was 16 847 for one candidate and 14 928 for the other. What percentage of the voters did not vote?

24   The total sum spent by a certain Town Council in 1969-70 was £825 000. Of this £440 000 was spent on education. What percentage of the total expenditure was spent on education?

25a   Show that $(1\frac{1}{2}-\frac{1}{3})+\frac{1}{4}$ has the same value as $1\frac{1}{2}-(\frac{1}{3}-\frac{1}{4})$.
  b   Show that $(1\frac{1}{2}\times\frac{1}{3})-\frac{1}{4}$ and $1\frac{1}{2}\times(\frac{1}{3}-\frac{1}{4})$ have different values.

26   Write out the $2\frac{2}{3}$ times table; $2\frac{2}{3}\times0 = ...,\ 2\frac{2}{3}\times1 = ...,\ 2\frac{2}{3}\times2 = ...,$ and so on up to $2\frac{2}{3}\times12 = ...$

Revision Exercise 3B

1 Insert one fraction between: *a* $1\frac{2}{5}$ and $1\frac{1}{2}$ *b* $\frac{3}{8}$ and $\frac{5}{12}$

2 Find the value of:

a $\frac{5}{8}$ of 2 right angles *b* $\frac{5}{12}$ of $8\frac{2}{5}$ *c* $\frac{3}{4}$ of £5·32

3 By noticing that $\frac{2}{3} = 1 - \frac{1}{3}, \frac{7}{8} = 1 - \frac{1}{8}$, and so on, evaluate:

a $\frac{9}{10}$ of 75 cm *b* $\frac{99}{100}$ of 3 litres *c* 99% of £37

4 Find the total cost of:
500 g cheese at 42p per kg; $\frac{1}{2}$ kg butter at 41p per kg;
500 g tea at 75p per kg;
10 slices bacon, averaging 35 g per slice, at 80p per kg

5 Find the total cost of:
$\frac{1}{2}$ litre white spirit at 31p per litre;
3 $\frac{1}{2}$-litre tins gloss paint at 58p per tin;
2 $\frac{1}{4}$-litre tins gloss paint at $31\frac{1}{2}$p per tin;
$1\frac{1}{4}$ litres emulsion paint at 72p per litre;
$\frac{3}{4}$ litre undercoat at 86p per litre

6 At a football match, $\frac{2}{3}$ of the spectators were in the stand, $\frac{1}{8}$ were in the enclosure, and 28 500 were on the terracing. How many spectators were at the match?

7 My savings increase every year by $\frac{1}{5}$ of what they amounted to at the beginning of that year. If my savings were £375 at the beginning of 1967, what were they at the beginning of 1970?

8 Divide the sum of $2\frac{1}{4}$ and $1\frac{7}{8}$ by their difference.

9 A litre of water weighs 1 kilogramme, and petrol weighs $\frac{7}{10}$ of the weight of the same volume of water. What is the weight of petrol in the tank of a car which holds 35 litres?

10 A housewife pays $82\frac{1}{2}$ pence for a joint of meat weighing $1\frac{1}{4}$ kilogrammes. What price per kg was charged?

11 A train leaves Edinburgh at 12.15 pm and arrives in Glasgow at 1.35 pm. The distance is 67 kilometres. What is the average speed (to nearest whole number) in kilometres per hour?

12 If $\square$ means 'square the first number and subtract the second' evaluate:

a $4 \square 2$ *b* $\frac{2}{3} \square \frac{1}{4}$ *c* $\frac{1}{2} \square \frac{1}{4}$ *d* $1\frac{1}{2} \square 1\frac{3}{8}$

13 The internal dimensions of a closed rectangular box are 16 cm by

$12\frac{1}{2}$ cm by 10 cm, and the external dimensions are 18 cm by 15 cm by 12 cm.

Calculate the volume of material in the walls of the box (including the top and bottom).

14 Find in the simplest form the ratio of:

a 85p to £2·55   **b**   5 cm 4 mm to 8 cm 1 mm

c 1 cm to 1 km

15 Two cuboids have dimensions 6 cm by 4 cm by 2 cm, and 8 cm by 6 cm by 4 cm respectively. Calculate the ratios of:

a the total lengths of their edges   **b**   the total areas of their faces

c their volumes

16 Express the following as fractions of 1 hour, and as percentages of 1 hour:

a 20 minutes **b** 12 minutes   **c** $7\frac{1}{2}$ minutes   **d** $\frac{3}{4}$ minute

17 In a bag of fertilizer, 5% of the weight is phosphate, and 24% is nitrogen. What weights of phosphate and nitrogen are there in 650 kilogrammes?

18 A dealer buys eggs at £1·92 per 12-dozen case. He wants to make a profit of $12\frac{1}{2}$% on what they cost him. What price should he charge per dozen?

19 In an arithmetic test a pupil scored 45 marks out of 120. What percentage is this? How many more marks would he have required in order to pass if the pass mark was:

a 50%   **b** 40%   **c** 45%?

20 A boy withdrew £4·50 from his Savings Bank account before going on holiday. If the sum remaining to his credit in the bank was £1·75, express the sum withdrawn as a percentage of his original savings.

21 Caustic soda contains, by weight, 40% oxygen, $2\frac{1}{2}$% hydrogen, and the rest sodium. How many kg of sodium can be obtained from 1200 kg of caustic soda?

22 A scientist knows that his measurements are liable to have an error of 4%. If he finds a measurement to be 5 cm, between what values does the correct length lie?

23 Every molecule of water is made up of 2 atoms of hydrogen and 1 atom of oxygen. The weight of an atom of hydrogen is 1 unit and the weight of an atom of oxygen is 16 units. What percentages of the weight of water are due to hydrogen and to oxygen?

24 Figure 5 shows the relative standard sizes of sheets of paper. The large sheet is folded in half to give the part A1, then the right-hand half is halved again to give A2, and so on. Write down the ratios:

a A2 : A1   *b* A5 : A3   *c* A7 : A1
d A2 : A6   *e* A5 : A7   *f* A6 : A1

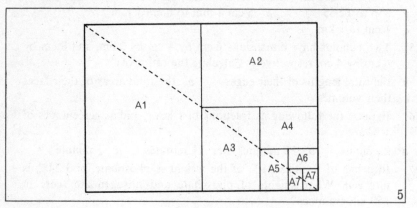

5

25 The associative law of *addition* can be stated
$$(a+b)+c = a+(b+c)$$
and *of multiplication* $(a \times b) \times c = a \times (b \times c)$.
Verify that the law is true in both cases when $a = \frac{1}{2}$, $b = \frac{1}{3}$, $c = \frac{3}{4}$.

26 Copy and complete the following:
a $\frac{3}{5} \times ? = \frac{1}{2}$ *b* $1\frac{1}{3} \div ? = \frac{5}{9}$ *c* $2\frac{3}{4} - ? = 1\frac{1}{3}$

27 Find the fraction x in each of the following:
a $4x = 5$ *b* $\frac{2}{3}x = \frac{1}{2}$ *c* $x - \frac{3}{8} = 1\frac{1}{4}$ *d* $\frac{3}{8}x = \frac{5}{16}$

28 P is the sequence $0, \frac{1}{2}, \frac{2}{3}, \frac{3}{4}, \frac{4}{5}, \ldots$
and Q is the sequence— $1, \frac{1}{2}, \frac{1}{3}, \frac{1}{4}, \frac{1}{5}, \ldots$
Show that:

a the difference between any of the first 6 terms of P and the preceding term is a term of Q.

b the difference between any of the first 6 terms of Q and the succeeding term is also a term of Q.

'True-False' Revision Exercise

Answer the following questions by writing T if the statement is true, F if it is false.

1 $38\frac{1}{2}p \times 100 = £38 \cdot 50$.

2      $3.5 \text{ kg} - 250 \text{ g} = 100 \text{ g}$.

3   10 pylons stand in a straight line, spaced 100 m apart, so the distance from the first to the last is 1 km.

4   The volume of a brick 23 cm by 10 cm by 7·5 cm is 1725 cm³.

5   The number of small cuboids each 4 cm by 3 cm by 2 cm that will completely fill a box 24 cm by 18 cm by 12 cm is 216.

6   There are 1 million square metres in a square kilometre.

7   129 694 is divisible by 4.

8   279 316 is divisible by 9.

9   If S is the set of square numbers and C is the set of cubic numbers then 1 and 64 are both members of $S \cap C$.

10   The next term of the sequence 5, 10, 20, 40, ... is 100.

11     $17 \times 15 \times 0 = 255$.

12   101 is a prime number.

13     $18^2 = 3^2 \times 6^2$.

14     $17^2 = 9^2 + 8^2$.

15   The operation of 'division' is commutative.

16   If a number is divisible by 3 and also by 4, it is divisible by 12.

17     $7\frac{3}{4} + 8\frac{1}{3} = 15\frac{4}{7}$.

18     $\frac{7}{12}$ lies between $\frac{5}{8}$ and $\frac{2}{3}$ on the number line.

19   53% of the pupils in the school are boys and 48% of the pupils are girls.

20     $13\frac{3}{4}$% of £320 is £44.

21   The product of 3 consecutive whole numbers is always divisible by 6.

22   The product of 3 consecutive odd whole numbers is always divisible by 9.

23   Jim is taller than Bert and Bert is taller than Tom, so Jim must be taller than Tom.

24   Betty has more money than Anne and Betty has more money than Jean, so Anne must have more than Jean.

25   A man's weekly pay of £20 was increased by 10% and later his new pay was decreased by 10%; so he finally gets the same pay as he had originally.

Answers

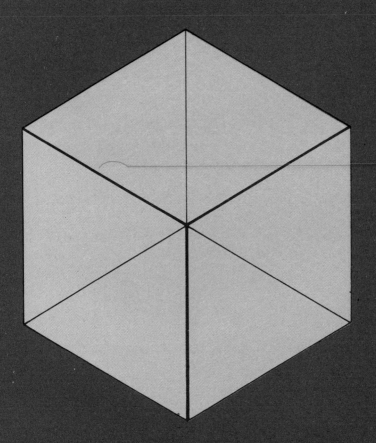

Answers

Algebra—Answers to Chapter 1

Page 4 Exercise 1A

$A = \{a, b, c, d, e, f\}$ $\qquad\qquad$ $B = \{11, 12, 13, 14\}$

$C = \{$Sun, Mon, Tue, Wed, Thu, Fri, Sat$\}$

$D = \{$red, amber, green$\}$ $\qquad\qquad$ $E = \{4, 6, 8\}$

$F = \{11, 13, 15, 17, 19\}$ $\quad$ $G = \{$Tue, Thu$\}$ $\quad$ $H = \{$Sep, Apr, Jun, Nov$\}$

$J = \{33, 53\}$ $\quad$ $K = \{0, 1, 2, 3, 4, 5, 6, 7\}$

$L = \{1, 2, 3, 4, 5, 6, 7, 8, 9, 10, 11, 12\}$

Page 4 Exercise 1B

$A = \{v, w, x, y, z\}$ $\qquad\qquad$ $B = \{98, 99, 100, 101, 102, 103, 104\}$

$C = \{\frac{1}{2}p, 1p, 2p, 5p, 10p, 50p\}$ $\qquad$ $D = \{10, 14, 20, 28\}$

$E = \{$H, E, I, C, S$\}$ $\qquad\qquad$ $F = \{21, 23, 25, 27, 29\}$

$G = \{1, 2, 3, 4, 6, 9, 12, 18, 36\}$ $\qquad$ $H = \{1972, 1976, 1980, 1984, 1988\}$

$I = \{0, 1, 4, 9, 16, 25\}$ $\qquad\qquad$ $J = \{234, 654\}$

$K = \{2, 3, 5, 7, 11, 13, 17\}$

$L = \{$red, orange, yellow, green, blue, indigo, violet$\}$

Page 5 Exercise 2A

1 set of first five letters of alphabet

2 set of days of week beginning with S

3 set of colours in traffic lights

4 set of seasons of the year

5 set of months of year beginning with J

6 set of capitals of France, England, Italy, Russia

7 set of metric weights in common use

8 set of British decimal silver coins

9 set of first six even whole numbers

10 set of whole numbers from 11 to 14 inclusive

11 set of operation signs used in arithmetic

12 set of numbers using numerals 1, 2, 3

Page 5 Exercise 2B

1 $\{864, 874, 884, 894, 904, 914, 924\}$

2a set of whole numbers $\qquad$ b set of odd numbers
 c set of whole numbers each multiplied by 3

3 {shapes with 3 sides from given list}, etc.

4 {4 pm, 3 am, noon} = {times from given list} etc.

5a $A = \{2, 4, 6, 8, 10\}$ $\quad$ b $B = \{2, 4\}$ $\quad$ c $C = \{1, 5\}$
 d $D = \{0, 5, 10, 15, 20\}$

6 $P = \{4, 6, 8, 10\}$ $Q = \{24, 48, 72, 96\}$ $S = \{1, 2, 3, 4\}$
   $T = \{6, 8, 10, 12, 14\}$

7 $\{258, 358, 458\}$ 8 $\{2, 3, 4, 5, 6, 7, 8, 9, 10, 11, 12\}$

Page 7 Exercise 3

1 starling, eagle, sparrow ∈ {birds}; giraffe, pony ∈ {animals}; shark ∈ {fish};
   apple, orange ∈ {fruit}; oak, chestnut ∈ {trees}

2a T   *b* T   *c* F   *d* T   *e* F   *f* F

3 $2, 4 \in \{1, 2, 3, 4\}$; $a \in \{a, b, c, d\}$; $6 \in \{6\}$; cm, km, ∈ {mm, cm, m, km};
   Mon, Fri ∈ {days of the week}; $4, 6 \in \{4, 5, 6\}$; $k \in \{k, l\}$; $Q, R \in \{Q, R, S, T\}$

4a $5 \in A$ *b* $5 \notin B$ *c* $15 \in B$ *d* $15 \in C$ *e* $300 \notin B$ *f* $300 \in C$

5a 3   *b* 5   *c* 1, 2, 4   *6a* F   *b* T   *c* T   *d* F   *e* T   *f* F

Page 8 Exercise 4

Empty sets are 1, 2, 3, 5, 7, 8, 9, 10, 12, 14, 16, 18.

Page 9 Exercise 5

1 $\{2, 4, 6\} = \{4, 2, 6\}$; $\{y, x\} = \{x, y\}$; $\{1, 3, 5, 7\}$ = {first four odd numbers};
   $\{1, 4, 9, 16\} = \{1 \times 1, 2 \times 2, 3 \times 3, 4 \times 4\}$;
   {vowels in English alphabet} = $\{u, e, i, o, a\}$

2 no   *3* $A = C, B = E$ and $D = F$ *4a* $Y = \{P, A, R, L, E\}$ *b* yes

Page 11 Exercise 6A

1a $\{A, C, E, F, J, K, M, P, Q\}$ *b* $\{A, K\}$ *c* $\{C, E, F, J, P\}$

   d $\{M, Q\}$ *e* $\{B, G, I, L, N, O\}$ *f* $\{B, D, G, L, R\}$ *g* $\{H, I\}$

   h $\{N, O\}$ *i* $\{D, H, R\}$ *j* ø or { }

2a $\{3, 5, 7\}$ *b* $\{3, 9\}$ *c* ø   *d* $\{7, 9\}$ *e* $\{5, 9\}$ *f* ø

3 $\{1\}, \{2\}, \{1, 2\}$, ø   *4* football—{forwards}, {full-backs}, etc.

5 {John}, {Mary}, {Betty}, {Mary, Betty}, etc.

6 $\{a, b\}, \{a, c\}, \{a, d\}, \{b, c\}, \{b, d,\} \{c, d\}$

7a T   *b* F   *c* T   *d* F   *e* T   *f* T   *g* T   *h* T

Page 12 Exercise 6B

1a $\{2, 4, 6\}$ *b* $\{3, 6\}$ *c* ø   *d* $\{1, 2, 3, 4, 6\}$ *e* $\{4, 5, 6\}$

2 $E \subset A, C \subset A, D \subset F, D \subset B, C \subset E$

3 e.g. {even numbers from 2 to 8}, {odd numbers from 1 to 7}, etc.

4a ø   *b* $\{3, 6\}$ *c* $\{5, 7\}$ *d* $\{4, 6, 8\}$

5a $\{p\}$, ø   *b* $\{p\}, \{q\}, \{p, q\}$, ø

   c $\{p\}, \{q\}, \{r\}, \{p, q\}, \{p, r\}, \{q, r\}, \{p, q, r\}$, ø

   d $\{p\}, \{q\}, \{r\}, \{s\}, \{p, q\}, \{p, r\}, \{p, s\}, \{q, r\}, \{q, s\}, \{r, s\}, \{p, q, r\},$
      $\{p, q, s\}, \{q, r, s\}, \{p, r, s\}, \{p, q, r, s\}$, ø.

6 1, 2; 2, 4; 3, 8; 4, 16; 5, 32

7a {cuboids}   *b* {cylinders, spheres}   *c* ø   *d*   {cuboids, cylinders, cones}

8 yes   **9** *a, d, e, f*

Page 14   Exercise 7

1 {letters of the alphabet}   **2** {domestic pets}   **3** {countries of Europe}

4 {odd numbers}   **5** {planets of the sun}   **6** {all metals}

7 {all vegetables}   **8** {pupils in your school}

9 {British silver coins}   **10** {all colours}

11 {odd numbers}, {whole numbers}

12 {vowels}, {letters of the alphabet}

13 {residents in Scotland}, {residents in Great Britain}

14 {all European cars}, {all cars in the world}

15 {numbers divisible by 3}, {natural numbers}

16 {even numbers}, {whole numbers}

17 {solids with flat surfaces only}, {solids with 8 corners}

18 {whole numbers between 50 and 60}, {natural numbers}

Page 15   Exercise 8A

1a   **b**   **c**

2a   **b**   **c**

3   **4a**   **b**   **c**

5a $A = \{1, 2, 3, 4\}$ **b** $B = \{3, 4, 5, 6, 7\}$ **c** $\{3, 4\}$
d $\{1, 2, 3, 4, 5, 6, 7\}$ **e** $\{ \ \}$

6a $C = \{p, q, r\}$ **b** $D = \{s, t, u, v\}$ **c** $\{ \ \}$
d $\{p, q, r, s, t, u, v\}$ **e** $\{w, x, y\}$

7a 6   **b** 7   **c** 3   **d** 3   **e** 4   **f** 4   **g** 14

Page 16   Exercise 8B

1 $E = \{0, 1, 4, 9, 16, 25, 36\}$
$A = \{16, 36\}$

2

3a

b

c

d

 $P = Q$

4

5a Fig. 14   **b** Fig. 15   **c** Fig. 16
d Fig. 13   **e** Fig. 15

Page 18   Exercise 9A

1 $A \cap B = \{7, 9\}$, $A \cap C = \{1, 3\}$

2 $D \cap F = \{2\}$, $D \cap W = \{2, 3, 5, 7, 11\}$

3 $P \cap Q = \{b, c, d\}$, $Q \cap R = \{d\}$, $R \cap S = \{f\} = S$,
$P \cap R = \{d, e, f\} = R$, $Q \cap S = \emptyset$

4a $A \cap B = \{3, 4\}$, $B \cap C = \{5\} = C$, $A \cap C = \emptyset$, $E \cap B = \{3, 4, 5\} = B$
b

$A \cap B$ $B \cap C = C$ $A \cap C = \emptyset$ $E \cap B = B$

5

6a $X = \{0, 2, 4, 6, 8\}$, $Y = \{1, 3, 5, 7\}$, $Z = \{3, 6, 9\}$
b $X \cap Y = \{\ \}$ or $\emptyset$, $Y \cap Z = \{3\}$, $X \cap Z = \{6\}$

7

8

9a   *b* 23   *10a*   *b* 4

Page 19   Exercise 9B

1a $A = \{3, 6, 9, 12, 15, 18\}$, $B = \{2, 4, 6, 8, 10, 12, 14, 16, 18\}$
  b $\{6, 12, 18\}$ *c* $\{6, 12, 18\}$ *d* yes
2a $Q = \{1, 2, 3, 4, 6, 9, 12, 18, 36\}$ *b* $\{1, 2, 3, 4, 6, 12\}$ *c* 12

3   *5a* $M = \{b, c, d, e\}$ *b* $S = \{a, b, d\}$
                               c $M \cap S = \{b, d\}$

6a   *b* 2   *7*   Ans. 2

8a $\{3\}$ *b* $\{p, r\}$
9a the set of pupils in your school whose surnames start with T
  b the set of pupils in your school who are 12 years old
  c the set of pupils in your town who are 12 years old and whose surnames
      start with T
  d the set of pupils in your school who are 12 years old and whose surnames
      start with T

10a $A \cap B = \{3, 5, 6\}$, $B \cap C = \{5, 6, 9\}$, $C \cap A = \{4, 5, 6\}$,
      $A \cap B \cap C = \{5, 6\}$

  b

Page 21   Exercise 10

1a T	*b* F	*c* F	*d* T	*e* T
f F	*g* T	*h* T	*i* T	*j* T

2a $A \cap B = \{3, 4, 5\}$ *b* $D \subset C, C \cap D = D$ *c* $F \cap G = \emptyset$
d $H = K, H \subset K, K \subset H, H \cap K = H = K$

3

    a   *b*   *c*   *d*

4 $X = Y, X \cap Y = X = Y$ *5* $B \subset A, A \cap B = B$
6 $a \in Y$ *7*

8 6   *9a* (ii)   *b* (iv)
                     c (i)   *d* (iii)
                     10 60

Algebra—Answers to Chapter 2

Page 24   Exercise 1A

1 T	*2* F	*3* T	*4* F	*5* T	*6* F
7 F	*8* T	*9* T	*10* F	*11* T	*12* F

Page 24   Exercise 1B

1 F	*2* T	*3* T	*4* T	*5* F	*6* T
7 T	*8* F	*9* F	*10* T	*11* F	*12* F

Page 26   Exercise 2

1 $3 + 5 = 8$ *2* $3 - 2 = 1$ *3* e.g. 1 ∈ {odd numbers}
4 e.g. 11 is greater than 10   *5* e.g. 2 ∈ {prime numbers}
6 e.g. 14 is a multiple of 7   *7* e.g. 20 is divisible by 10
8 e.g. two million is greater than one million

9 2	*10* 7	*11* 5	*12* 4	*13* e.g. 9
14 e.g. 2	*15* 10	*16* 6	*17* 36	*18* 1

19 {1, 3, 5, 15}   *20* {8}   *21* {1, 2, 3, 4, 5, 6, 7}
22 {15}   *23* {20}   *24* {4, 8, 12, 16, 20}
25 {6}   *26* {6, 7, 8, 9}

Page 27   Exercise 3

1 60	*2* 6	*3* 12	*4* 2	*5* 7	*6* 6

7a {3, 9, 15}   *b* {6, 12}   *c* {3, 6, 9, 12, 15}
 d {3, 6}   *e* {15}   *f* {3, 6, 9}
8a {4, 10}   *b* {7, 13}   *c* {10}
 d { } or ø   *e* {4, 7}   *f* {4, 7, 10, 13}

9a	{3}	*b*	{7, 9}	*c*	{1, 3, 5}
d	{1, 3, 5, 7}	*e*	{1, 3}	*f*	{ } or ø
10a	{8}	*b*	{6, 7, 8, 9, 10}	*c*	{1, 2, 3}
d	ø	*e*	{1, 4, 9}	*f*	{4}

Page 29   Exercise 4A

1	$x = 7$	*2*	$m = 6$	*3*	$y = 10$	*4*	$p = 9$
5	$x = 17$	*6*	$a = 5$	*7*	$z = 10$	*8*	$a = 11$
9	$x = 9$	*10*	none	*11*	$x = 18$	*12*	$x = 4$
13	$m = 49$	*14*	$y = 350$	*15*	none	*16*	$x = 27$
17	$x = 18$	*18*	$y = 54$	*19*	$m = 5$	*20*	$p = 8$
21	$p = 1$	*22*	$n = 11$	*23*	$x = 7$	*24*	none
25	$x = 3$	*26*	$y = 10$	*27*	$z = 25$		

Page 29   Exercise 4B

1	{1½}	*2*	{¼}	*3*	{3½}	*4*	{1½}	*5*	{3½}
6	{6¼}	*7*	{5½}	*8*	{ } or ø	*9*	{¾}	*10*	{13}
11	{ } or ø	*12*	{½}	*13*	{¼}	*14*	{⅓}	*15*	{3}
16	{¾}	*17*	{⅛}	*18*	{6}	*19*	{10}	*20*	{1000}
21	{½}	*22*	{9}	*23*	{2}	*24*	{17}		

Page 30   Exercise 5

1a rows: 0, 1, 2, 3; 1, 2, 3, 0; 2, 3, 0, 1; 3, 0, 1, 2
  b yes; symmetry about main diagonal
  c e.g. only the four numbers 0, 1, 2, 3, are needed; 0 is the identity element for addition

2a	$x = 2$	*b*	$x = 3$	*c*	$x = 1$	*d*	$x = 1$
e	$x = 2$	*f*	$x = 3$	*g*	$x = 3$	*h*	$x = 3$
i	$x = 2$						

3a back one or forward three   *b* forward or back two   *c* 1 or 3   *d* 0
4 rows: 0, 1, 2, 3, 4, 5; 1, 2, 3, 4, 5, 0; 2, 3, 4, 5, 0, 1; 3, 4, 5, 0, 1, 2; 4, 5, 0, 1, 2, 3; 5, 0, 1, 2, 3, 4

a	$x = 4$	*b*	$x = 2$	*c*	$x = 2$	*d*	$x = 3$
e	$x = 4$	*f*	$x = 4$				

5a	1	*b*	5	*c*	4	*d*	2	*e*	1
f	4	*g*	2	*h*	5	*i*	3. Subtraction is always possible,

6a	1	*b*	3	*c*	5	*d*	4	*e*	2
f	1	*g*	0	*h*	3	*i*	4		

Page 32   Exercise 6A

1	$x+4 = 11$	*2*	$4+y = 9$	*3*	$x-7 = 12$
4	$t-11 = 2$	*5*	$5+x = 14$	*6*	$x-3 = 8$

7 $x+4 = 10$ *8* $y-4 = 2$ *9* $m-4 = 6$

10 $q+q = 14$ *11* $x-y = 0$ *12* $n = 15+4$

13 $m = n+2$ *14* $x+y = 12$ *15* $p-q = 29$

16 $a = b+5$ *17* $a+b+35 = 360$ *18* $x+y+z = 180$

19 7, 5, 19, 13, 9, 11, 6, 6, 10, 7

Page 33   Exercise 6B

1 $30-x = 25; x = 5$ *2* $21+x = 30; x = 9$

3 $n+100 = 850; n = 750$ *4* $325-p = 84; p = 241$

5 $250+x = 320; x = 70$ *6* $a = b+5; a = 50$

7 $p = q-20; q = 100$ *8* $r+5+15 = 80; r = 60$

9 $x+5 = 63; 58$ kg   *10* $x+y = 100; x = 55$

11 $q = 4+x$ *a* $q = 14$ *b* $x = 32$ *12a* 21   *b* 8   *c* 12

Page 35   Exercise 7

1 $y = 1+x; y = 2; y = 3; y$: 2, 3, 4, 5, 6, 7

2a $y = x-1$ *b* 0, 1, 2, 3, 4, 5

3a $y = 2x$ *b* 2, 4, 6, 8, 10, 12

4b $y = x+5$ *c* y: 6, 7, 8, 9, 10, 11

5b $y = x+8$ *c* y: 9, 10, 11, 12, 13, 14

6a 20 km   *b* $10+x+y = 20$ *c* $x+y = 10$ *d* y: 9, 8, 7, 6, 5, 4

Algebra—Answers to Chapter 3

Page 37   Exercise 1

1 3×2 *2* 2×3 *3* 4×8 *4* 5×7 *5* 4×1

6 2×9 *7* 3×3 *8* $2\times a$ *9* $3\times n$ *10* $5\times y$

11 $4\times k$ *12* $2\times c$ *13* $10\times p$ *14* $15\times r$ *15* $12\times x$

16 $6\times b$ *17* $5+5$ *18* $8+8+8$

19 $2+2+2+2+2+2+2$ *20* $4+4+4+4+4$

21 $0+0+0+0+0+0$ *22* $9+9+9+9$

23 $w+w+w+w+w$ *24* $y+y+y$

25 $n+n+n+n$ *26* x *27* y

28 $z+z+z+z+z+z+z+z+z+z$

Page 38   Exercise 2

1 6   *2* 6   *3* 3   *4* 5   *5* 1   *6* 0

7 2   *8* 5   *9* 5   *10* 6   *11* 4   *12* 3

13 5   *14* 1   *15* 4   *16* 2   *17* 5   *18* 3

Page 39   Exercise 3A

1 $3m$	**2** $7p$	**3** $2x$	**4** $12p$	**5** y	**6** $8c$
7 y^2	**8** $3x$	**9** $9r$	**10** $10t$	**11** $12h$	**12** $15k$
13 c^2	**14** xy	**15** pq	**16** m^2		

17 (i) $2x$ cm² (ii) pq cm² (iii) y^2 cm²   **18** 4   **19** 2

20 9	**21** 3	**22** 7	**23** 12	**24** 1	**25** 2
26 3	**27** 9	**28** 1	**29** 7	**30** 15	**31** 42
32 12	**33** 1				

34a 8	*b* 4	*c* 7	*d* 9	*e* 15	*f* 4	*g* 2	*h* 4
35a 3	*b* 1	*c* 2	*d* 8	*e* 2	*f* 5	*g* 13	*h* 3
36a 4	*b* 0	*c* 11	*d* 20	*e* 10	*f* 0	*g* 5	*h* 8

Page 40   Exercise 3B

1 (i) $3x$ cm   (ii) $4y$ cm   (iii) $6z$ cm   **2a** $12t$ cm   *b* $t^2, 6t^2$ cm²   *c* 120 cm

3a $12x$ cm   *b* $12x$ cm   *c* $5x^2$ cm²   *d* 24 cm

4a 7   *b* 20   *c* 25   *d* 11   *e* 20   *f* 0   *g* 0   *h* 36

5a 1   *b* 12   *c* 13   *d* 0

6a 2, 4, 6   *b* 9, 10, 11   *c* 0, 3, 6   *d* 8, 5, 0   **7** 18, 0

8a $7k$ days   *b* $100p$ cm   *c* $100q$ pence   *d* $12x$ months   *e* $1000y$ g   *f* $1000z$ ml

9a 250p   *b* $5x$ p   *c* $50x$ p   *d* x^2 p

10a rows: $0, 0, 0$; x, y, z; $2x, 2y, 2z$ *b* rows: i^2, ij, ik; ji, j^2, jk; ki, kj, k^2

11 32   **12** 3, 4, 5

Page 43   Exercise 4A

1 90	**2** 3800	**3** 700	**4** 1900	**5** 2600	**6** 1500
7 $6a$	**8** $4p$	**9** $40r$	**10** $5ab$	**11** $3mn$	**12** $8t$
13 $6x$	**14** x	**15** $7gh$	**16** a^2b	**17** c^3	**18** $15w$

19 $8uv$ **20** $8xy$ **21** $10mn$ **22** (i) $8ax$ cm³ (ii) z^3 cm³ (iii) p^2q cm³

23a 18   *b* 30   *c* 18   *d* 60   *e* 900

24a 120   *b* 80   *c* 32   *d* 44

25a rows: $x, 2y, 3z$; $2x, 4y, 6z$; $3x, 6y, 9z$
   b rows: $x^2, 2xy, 3xz$; $2xy, 4y^2, 6yz$; $3xz, 6yz, 9z^2$ **26a** 280   *b* 900

Page 45   Exercise 4B

1 $16mn$	**2** $21uv$	**3** $6x^2$	**4** $3g^2h$
5 a^2bc	**6** a^2b^2	**7** $6hk$	**8** $40xy$
9 $9abc$	**10** $90xy$	**11** $24abc$	**12** 0

13a 9   *b* 0   *c* 15   *d* 90   *e* 23   *f* 73

14a 21   *b* 42   *c* 48   *d* 190   *e* 45   *f* 100

15a 23   *b* 33   *c* 1080

16a rows: 0, 0, 0, 0; 0, 1, 2, 3; 0, 2, 0, 2; 0, 3, 2, 1

   b yes   *c* (*1*) {1, 3}   (*2*) {0}   (*3*) {2}   (*4*) {3}   (*5*) {0, 2}   (*6*) ø

Page 46   Exercise 5A

1 T	*2* T	*3* T	*4* F	*5* T	*6* F
7 2	*8* 2	*9* 5	*10* 4	*11* 10	*12* 3
13 0	*14* 1	*15* 2	*16* 1	*17* 2	*18* 0
19 6	*20* 4	*21* 10	*22* 3	*23* 4	*24* 5
25 $2\frac{1}{2}$	*26* $1\frac{1}{2}$	*27* 8	*28* $4\frac{1}{2}$	*29* $\frac{1}{3}$	*30* $\frac{2}{3}$

Page 47   Exercise 5B

1 T	*2* T	*3* F	*4* F	*5* T	*6* T
7 3	*8* 6	*9* 4	*10* 1	*11* 4	*12* 8
13 5	*14* 6	*15* 3	*16* 5	*17* 4	*18* 3
19 0	*20* 7	*21* 19	*22* 18	*23* 25	*24* 9
25 $2\frac{1}{2}$	*26* $7\frac{1}{2}$	*27* $13\frac{1}{2}$	*28* 24	*29* 4	*30* 6

31a {0, 4}   *b* {0}   *c* {1}   *d* {1, 5}

32 $x = 1, y = 3; x = 3, y = 2$

Page 48   Exercise 6A

1 $5x$	*2* $7k$	*3* $10a$	*4* $8m$	*5* $11h$	*6* $15u$
7 $10c$	*8* $10t$	*9* $6y$	*10* $13b$	*11* a	*12* c
13 0	*14* $7n$	*15* $8x^2$	*16* a^2	*17* $13ab$	*18* $5x^3$
19a $6w$	*b* none	*c* $7x$	*d* $10x$	*e* none	*f* $11a$
20 8	*21* 8	*22* 1	*23* 1	*24* 3	*25* 0
26 $9x$	*27* $8p$	*28* $20m$	*29* $6h$	*30* $9w$	*31* $11y$
32 $8x$	*33* 0	*34* z	*35* $12x^2$	*36* $2y^2$	*37* $9z^2$

Page 49   Exercise 6B

1 $15x$	*2* $5y$	*3* $8a$	*4* $5b$
5 0	*6* not possible	*7* $8x + 6y$	*8* $5x + 4y$
9 $8x + 4y$	*10* $7a$	*11* $8a + 5b$	*12* $9c + d$
13 $5x^2 + 2y^2$	*14* $9x^2 + y^2$	*15* $3a^2$	

16a $3p$ cm   *b* $5n$ cm   *c* $10m$ cm   *d* $10t$ cm   *17* $5a$ cm, 8

18 $10n$ cm, 9·5   *19a* $8p$ cm   *b* 40 cm   *c* 4

20a $10x$ m   *b* $20x$ m   *c* 5; 35 m, 15 m   *21a* $30y$ cm   *b* 3   *c*   24 cm

22(i) $2x + 2y + 2z$ m   (ii) $2a + 2b + 2c$ m   (iii) $6p + 4q$ m

23b $2x + 6$ m   *c* $2x + 6 = 16$ *d*   5; 15 m^2

24a 10*d* m   *b* 10*d* = 80; *d* = 8   *c* 56 m

25a *x* + 15   *b* 2*x* + 15   *c* 2*x* + 15 = 137; *x* = 61; 76

Algebra—Answers to Revision Exercises

Page 53   Revision Exercise 1A

1a {January, February, May, July}   *b* {2, 3, 5, 7, 11, 13, 17, 19, 23, 29}
c {L, M, N, O, P, Q}   *d* {21, 23, 25, 27, 29}

2a the set of months beginning with the letter J
b the set of known planets of the Sun
c the set of solids with one vertex
d the set of whole numbers formed by the digits 2, 3 without repetition of the digits

3   *a* ∈ {*a*, *b*, *c*}, *a* ∈ {vowels of English alphabet}, green ∈ {colours},
green ∈ {red, yellow, green}, 12 ∈ {even numbers}, 12 ∈ {10, 11, 12, 13},
b ∈ {*a*, *b*, *c*}, 13 ∈ {10, 11, 12, 13}

4a T   *b* F   *c* F   *d* T   *e* T

5a 3 ∈ *W*   *b* ∅   *c* *x* ∉ *A*   *d* *S* ⊂ *T*   *e* *P* = *Q*

6a {2, 3, 5, 7}   *b* {1, 3, 5, 7, 9}   *c* {1, 2, 5, 7, 10}   *d* {3, 6, 9}

7a {1, 2, 3}, {1, 2, 4}, {1, 2, 5}, {1, 3, 5}, {1, 4, 5}, {1, 3, 4} {2, 3, 4}, {2, 3, 5},
{2, 4, 5}, {3, 4, 5}.

b 10   *c* None of the intersections is empty.

8   *a* (ii),   *b* (iv),   *c* (i),   *d* (iii)
Shade *B* in (i); shade *A* in (ii); no shading for (iii); shade *A*, *B* in (iv).

9a {2, 3}   *b* ∅   *c* {1, 2, 3}   *d* ∅

10a {*b*, *c*}   *c* *f* and *g*

11a *A* ∩ *B* = {3, 4}, *B* ∩ *C* = {4}, *A* ∩ *C* = {2, 4}
b *A* ∩ *B* ∩ *C* = {4}

12 *a*, *c* and *e* are true.

Page 55   Revision Exercise 1B

1a {1, 2, 3, 4, 5, 6, 7, 8, 9, 0}   *b* {1984, 1988}
d {touch, sight, smell, hearing, taste}

2a {8, 10, 12, 14, 15, 16}   *b* {9, 12}   *c* {5, 7, 11, 13}   *d* {1, 2}

3a {*ab*, *ac*, *ad*, *bc*, *bd*, *cd*}
b {*abc*, *abd*, *bcd*, *acd*}   *c* {*abcd*}

4a T   *b* F   *c* T

5 2 ∈ {even numbers}, 2 ∈ {first four whole numbers}, elephant ∈ {animals}, Paris ∈ {capitals of Europe}, 3 ∈ {odd numbers}, 3 ∈ {first four whole numbers}, Dublin ∈ {capitals of Europe}

6 *b* and *c*

7a {1, 3, 5, 7, 9}   *b* {2, 4, 6, 8}   *c* {1, 3, 7}   *d* {1, 2, 4, 5}   *e* {3, 5, 6}

8a {2, 3, 5, 7}   *b* {1, 3, 5, 7, 9}   *c* {1, 2, 3, 4, 6, 12}   *d* {3, 5, 7}   *e* {1, 3}

9a   *b*   *c*   *d*

10a The set of yellow roses in the shop
   b There are no yellow roses in this shop.

11a (ii)   *b* (i)   *c* (iv)   *d* (iii)

12 $B \cap C = \{10\}$, $A \cap C = \{5, 15\}$; $A \cap B = \emptyset$

13

      yes

14 *b*, *c* and *d* are true.

$Q \subset P$ $P \cap R = \emptyset$ $S \subset P$

15a {2, 3, 5}   *b* {2, 3, 8}   **16** $X \cap Z = X$
   c {2, 3, 7, 9}   *d* {2, 3}

Page 57   Revision Exercise 2A

1a T   *b* F   *c* T   *d* F   *e* T   *f* T

2a T   *b* F   *c* F   *d* T   *e* T   *f* T

3a {8}   *b* {5}   *c* {4}   *d* {5}   *e* {1, 2, 3, 4}
   f {1, 2, 3}

4a {1, 2, 3, 4, 6, 12}   *b* {1, 2, 3, 6}   *c* {0, 1, 2, 3, 4}
   d {23, 29}   *e* {3, 5, 7, 9}   *f* {9}

5a $x = 5$ *b* $x = 19$ *c* $y = 23$ *d* none
e $z = 36$ *f* $m = 4$ *g* $n = 3$ *h* $x = 9$ *i* $x = 9$

7a 1   *b* 2   *c* 0   *d* 2   *e* 3   *f* 12

8a $t + 10 = 24$ *b* $5 - x = 1$ *c* $y + 7 = 35$
d $x \times x = 169$ *e* $a - 8 = b$ *f* $15 + t = 21$

Page 59   Revision Exercise 2B

1a F   *b* F   *c* T   *d* T   *e* T   *f* T

2a {8}   *b* {8, 10}   *c* {6, 8, 10}   *d* ø   *e* {8}   *f* {4, 8}

3a 2   *b* 19   *c* 4   *d* ø   *e* 4   *f* 100
g ø   *h* every member of N *i* 15

4a {1, 2, 3}   *b* {6, 7}   *c* {9, 10, 1, 2, 3, 4}
d {1, 2, 3}   *e* {1, 2}   *f* {9}

5a c   *b* c   *c* e, a, b, c

6a 11   *b* 12   *c* 13, 14, 15, ..., 20

7a {(4, 8), (3, 9), (2, 10)}   *b* {(4, 10)}   *c* {(1, 7)}
d {(4, 8), (4, 9), (4, 10)}

8a {(8, 4), (7, 3), (6, 2), (5, 1)}
b {(1, 1), (1, 2), (2, 1), (1, 3), (3, 1), (1, 4), (4, 1), (1, 5), (5, 1), (1, 6), (6, 1), (1, 7), (7, 1), (1, 8), (8, 1)}

Page 60   Revision Exercise 3A

1a $14k$ *b* $8c$ *c* ab *d* a^2 *e* b^3

2 $48, 4x, pq, y^2, 12a$ cm^2 *3* $28, 8 + 2x, 2p + 2q, 4y, 2a + 24$ cm

4a 10   *b* 12   *c* 1   *d* 6

5a 17   *b* 14   *c* 32   *d* 50   *e* 125

6a $28a$ *b* $5xy$ *c* $6pq$ *d* $12k^2$

7a 11   *b* 10   *c* 0   *d* 49

8a T   *b* T   *c* F   *d* T

9a 4   *b* 12   *c* 9   *d* 9   *e* 3   *f* 5   *g* 8   *h* 0   *i* 11

10a $12x$ *b* $9y$ *c* $16z$ *d* $2ab$ *e* $5c$ *f* $13d^2$
g $6k$ *h* $12q$ *i* 0   *j* —   *k* —   *l* $8a - 2b$

11a $1\frac{1}{2}$ *b* $\frac{1}{2}$ *c* $4\frac{1}{2}$ *d* $1\frac{1}{2}$ *e* $\frac{1}{2}$ *f* $2\frac{1}{2}$

12 $24, 12a, 4ab, abc$ cm^3 *13* $52, 14a + 24, 8a + 8b + 2ab, 2ab + 2bc + 2ac$ cm^2

14a {1, 3}   *b* {1, 2, 3, 4}   *c* {3}   *d* ø   *e* {1}   *f* {4}

15a rows: $6x, 7x, 8x$; $5x, 6x, 7x$; $4x, 5x, 6x$
b rows: $5x^2, 10xy, 15xz$; $4xy, 8y^2, 12yz$; $3xz, 6yz, 9z^2$

Page 62   Revision Exercise 3B

1a 19   *b* 0   *c* 108   *d* 6   *e* 432

2a 60 cm   *b* 150 cm^2 *c* 125 cm^3

3a $12x$ m   *b* $6x^2$ m^2 *c* x^3 m^3

4a 0, 8, 16, 32 *b* 0, 4, 16, 64
 c 12, 16, 20, 28 *d* 100, 96, 84, 36

5a $100x$ p *b* $1000y$ m *c* $24z$ h *6a* £$8x$ *b* £y^2

7a 12 *b* 15 *c* 1 *d* none *e* 40 *f* 4 *g* 24 *h* 5

8a $10mn$ *b* $21kn$ *c* $12x^2$ *d* $4abc$ *e* $5p^2q$ *f* $24abc$

9a 0 *b* 14 *c* 36 *d* 77

10a 4 *b* 5 *c* 24 *d* 12 *e* 7 *f* 0
  g none *h* none *i* 25 *11a* $7\frac{1}{2}$ *b* $\frac{1}{2}$ *c* $2\frac{1}{2}$

12a $4a$ *b* $8b$ *c* — *d* x^2
  e $5x^2 + y^2$ *f* $6x^2$ *g* $5x + 5y$ *h* $2a + b$
  i $7x^2$ *j* $12ab$ *k* — *l* $10x + 1$

13a xyz cm³ *b* $4x + 4y + 4z$, or $4(x + y + z)$ cm
  c $2xy + 2yz + 2xz$, or $2(xy + yz + xz)$ cm²

14 18 m, 162 m² *15* $40n$, 20 *16* $133\frac{1}{3}$ cm, $66\frac{2}{3}$ cm

17a {4} *b* {2}, yes, symmetry of table

18a 4, 6, 8, 6, 9, 12, 8, 12, 16, yes *b* 14, 14, no, 28, 26

19a rows, 0, 1, 2, 3, 4; 1, 0, 1, 2, 3; 2, 1, 0, 1, 2; 3, 2, 1, 0, 1; 4, 3, 2, 1, 0; yes
  b 0, 2, no

20 3, 2, no.

Geometry—Answers to Chapter 1

Page 67 Exercise 1

1 matchbox, football, lamp, tin, die, ice-cream cone, house, pyramid

2 *object* matchbox; football; tin; die; ice-cream cone; pyramid
  shape cuboid; sphere; cylinder; cube; cone; pyramid

4a 8 *b* 12 *c* 6

5 cuboid: 6, 8, 12 cube: 6, 8, 12 pyramid: 5, 5, 8
  sphere: 1, 0, 0 cone: 2, 1, 1 cylinder: 3, 0, 2

6a 8, 12, 6 *b* walls, floor, roof or ceiling

7a (i), (v) *b* (ii), (iv) *8a* no *b* yes *c* yes

9 gate, aircraft, wheel, church, 50p coin, apple, door, television set,
  crescent moon, cup and saucer

Page 69 Exercise 2

1a cone, cube, pyramid, cuboid, cylinder
  b cube, cuboid *c* sphere, cone, pyramid, possibly cylinder

2a cuboid *c* 6 *d* 2 *e* 22 cm by 7 cm, 11 cm by 7 cm *3* 3

4a PQ, SR, DC *b* QR, PS, AD
  c AP, BQ, CR, DS *d* They are hidden from view. *6b* 6

8a AD, EH, FG *b* DH, CG, BF *c* AB, EF, HG, DC

9 no *10a* yes *b* no *11a* (i), (ii), (iii), (v) *b* (iv)

Page 72   Exercise 3

1a 4   *b* 4   *c* 4
d 96 cm needed. $(4 \times 5) + (4 \times 6) + (4 \times 13)$, or $4 \times (5 + 6 + 13)$ cm

2 48 cm   *3* 212 cm   *4a* 124 cm   *b* 98 cm   *c* 222 cm

5a yes   *6b* 72 cm   *7a* yes, a cube   *b* yes, a cuboid   *c* no

Page 75   Exercise 4

2a (i) yes (ii) no   *5a* (i) yes (ii) no   *6* (iii), (iv), (vii), (viii), (ix)

Page 77   Exercise 5

1a 12 cm by 5 cm, 5 cm by 20 cm, 20 cm by 12 cm

3a yes   *4* 4   *5a* 4   *b* 4   *c* 24   *6a* 4   *b* 2

Page 79   Exercise 6

3 4   *4a* 4   *b* 8   *5a* 2   *b* 4

6a (i), (ii)   *b* (ii)   *c* (i), (ii)

9a J   *b* I, K   *c* D, L, M

Page 82   Exercise 7

1 (ii) is red; (iii) is blue and red   *2a* yes   *b* no   *4* (v), (vii), (ix)

5 (vi), (viii)   *6a* N, H, Z, I, S, X   *b* A, V, H, Y, T, M, I, W, X

Geometry—Answers to Chapter 2

Page 85   Exercise 1

1 2, 2, 2, 1, 2   *2* 2, 2, 4, 4   *3* 4, 4, 8   *4* 2, 1, 2, 1, 4, 2, 2, 4, 1, 2

Page 87   Exercise 2

5a 3 o'clock, 9 o'clock   *c* yes   *d* 6 o'clock   *7* in a vertical direction

8 builder, bricklayer   *9* horizontal   *11a* perpendicular   *12* 3

Page 88   Exercise 2B

1b between 0454 and 0455 hours

2b between 0405 and 0406 hours; between 0438 and 0439 hours

3 Together they make up a straight angle (test on a straight edge).

4a T   *b* T   *c* F   *d* F   *e* T

Page 89   Exercise 3

3a a right angle   *b* two right angles, or one straight angle

c three right angles   *d* four right angles   *4* 10   *7* 18

8a 90   *b* 30   *c* 60   *d* 120   *e* 180
9a 30   *b* 60   *c* 120   *d* 90
 e 120   *f* 150   *g* 90   *h* 180
10a $\frac{1}{4}$ *b* $\frac{1}{8}$ *c* $\frac{1}{2}$ *d* $\frac{3}{4}$

Page 91   Exercise 4

7b 210°   *10a* 72   *b* 144   *c* 144   *d* 360
11a 30   *b* 180   *c* 720
12a right   *b* straight   *c* complete turn   *d* vertex   *e* 90   *f* 180

Page 94   Exercise 4B

2b 230°, 590°, etc.
3 1, 2, 3, 4, 5, 6, 8, 9, 10, 12, 15, 18, 20, 24, 30, 36,
    40, 45, 60, 72, 90, 120, 180, 360
4 5760° per second   *5a* 1°   *b* 92°

Page 95   Exercise 5

1a acute   *b* obtuse   *c* acute   *d* obtuse   *e* acute
2 acute: 10°, 19°, 89°, 1°, 77°   obtuse: 160°, 100°, 91°, 99°, 179°
   right: 90°   straight: 180°

3a right   *b* obtuse   *c* right   *d* acute
 e straight *f* obtuse   *g* obtuse   *h* acute
5a T   *b* T   *c* F   *d* T   *e* T   *f* T
6a angles AED, BEC, BAE, ABE, ADE, DCE, BCE, CDE, ADC, BCD
 b angles AEB, DEC, DAB, ABC, DAE, EBC

Page 96   Exercise 5B

1 a right angle

3a obtuse   *b* neither   *c* obtuse
 d neither   *e* acute   *f* obtuse

Page 97   Exercise 6

3a 90°   *b* 180°   *c* 135°   *d* 90°
4a east, west   *b* north-east, south-west
5a 90°   *b* 45°   *c* 135°
6a south   *b* east, west
7a $11\frac{1}{4}$°   *8a* 1, 4   *b* 2, 3; 4, 5; 5, 1   *c* 45°   *d* 135°

Page 99   Exercise 7

1a 090°   *b* 180°   *c* 270°   *d* 045°   *e* 135°

2

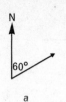

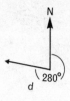

a   b   c   d   e

3 85°   *4* 107°   *5* 47°

Page 99   Exercise 7B

1 135°   *2* 265°   *3* 020°   *4* 180°   *5* E by N

Page 100   Exercise 8

1 180°

2a 55   *b* 108   *c* 132   *d* 81   *e* 180   *f* decreases by 10

3a 0   *b* 180   *c* 90   *4* 170°, 80°, 1°, 92°, 90°

5a ∠CBE   *b* ∠BCE

6a angles TQP, PQS; PQS, SQR; SQR, RQT; RQT, TQP
 b 115°, 65°, 115°

7 $x+z = 180, z+y = 180, y+u = 180, u+x = 180$

8a 180°   *b* They make a straight line.

Page 101   Exercise 9

1 90   *2a* 65   *b* 28   *c* $x = 90$

3 80°, 65°, 1°, 40°, 24°   *4a* 45°   *b* 30°, 60°

Page 102   Exercise 10

2a increases by 10°   *b* increases by x°

3 It lies along OA.   *4* They are equal.

5 When two lines cross you always get two pairs of vertically opposite angles.

Page 103   Exercise 10B

1 133, 53, 60, 63, 60, 60   *2* yes, *b*

3a 230°   *b* 130°   *c* They are supplementary.

4 $x° + y° = 180°$; also ∠BOD $+ y° = 180°$, so ∠BOD $= x°$

5

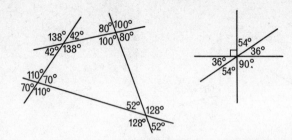

7a 360°   b 120°   c 50°, 70°, 60°   d 360
e (180 − a)°, (180 − b)°, (180 − c)°; 180°

Page 107 Exercise 11

1a 46 m   b 55 m   c 43 m   2 100 km, 143°

3 75 km, 307°   4 500 km, 227°   5 675 km, 061°

6a 140 m   b 49°

Page 108 Exercise 12

1 30 km   2 21 km   3 596 km   4 7·9 km   6 103 m   7 127 m

Geometry—Answers to Chapter 3

Page 111 Exercise 1

4a B   b C   c B(6, 4), C(1, 6), D(0, 3), E(5, 0), F(2, 4), G(5, 5)
5

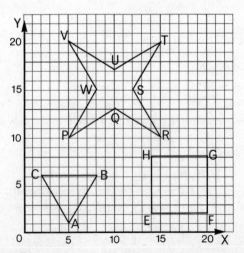

6   A(2, 2), B(1, 5), C(2, 6), D(3, 5);   E(1, 8), F(4, 8), G(7, 11), H(4, 11);
      J(4, 5), K(7, 6), L(8, 9);   M(6, 1), N(9, 3), P(12, 1), Q(9, 4);
      S(10, 7), T(12, 5), U(16, 9), V(14, 11)

8a   (2, 6), (4, 7), (5, 1)   *b*   Capital, Eastview, Northpoint
  c   90 km, 50 km   *11c*   (10, 6)   *12c*   (15, 13)
13a   2, 0, 100   *b*   7, 8, 1   *c*   (12, 12), (15, 15), (18, 18), (21, 21)

Page 113   Exercise 2A

2b   yes   *c*   (2, 2), (5, 5)   *3b* $a = b$
5b   yes, yes, yes   *c*   an infinite number
6b   yes, yes, yes   *c*   an infinite number
7   the point (5, 8)   *8*   at the point (3, 5)
9a   B(9, 4), D(4, 9)   *b*   E(14, 14)   *10b*   yes   *c*   3
11a   A(3, 1), B(6, 2), C(9, 3)   *b*   (12, 4), (15, 5), (18, 6)
  c   yes, yes, no, yes   *d*   36   *e*   8   *f*   (2)

Page 116   Exercise 2B

1b   (*1*)   (4, 4), (11, 11)   (*2*)   (8, 4)   (*3*)   (8, 0), (8, 4)
2a   four   *b*   (4, 2), (6, 2), (8, 2), (4, 4), (8, 4), (4, 6), (6, 6), (8, 6)
3a   (9, 6), (6, 12)   *b*   (12, 18)   *c*   (3, 0)   *d*   162 square units
4a   (10, 10), (12, 12), (14, 14), (16, 16)   *c* $r = s$; $r, s \in$ {even numbers}
  d   K, N   *e*   (3)
5 $Q \subset T$ *6b*   (5, 3), (10, 3), (10, 8), (5, 8)   *c*   ø, ø
7b   *b* is less than 10   *c*   above the line   *8b*   no; yes; no
9b   the interior of the rectangle bounded by the lines $x = 4$, $x = 8$,
      $y = 2$ and $y = 10$
10a   (2, 3)   *b*   (2, 7), (2, 10)   *c*   no   *d*   The *x*-coordinate is 2.
  e $x = 2$ and *y* is greater than 2.
11a   (4, 3), (8, 6), (12, 9)   *b*   (16, 12), (20, 15), (24, 18)
  c   no, no, yes, no   *d*   16   *e*   18   *f*   (3)

Geometry—Answers to Revision Exercises

Page 122   Revision Exercise 1A

3a   sphere   *b*   cuboid, cube, pyramid   *c*   cylinder, cone
4a   sphere   *b*   cone   *c*   pyramid
5   HG = DC = AB = 4 cm; BC = AD = EH = 3 cm;
      FB = EA = HD = 5 cm
6   {EF, HG, DC, AB}, {FG, BC, AD, EH}, {GC, FB, EA, HD}
7   60 cm   *8*   3 cm   *10*   two, 16 cm by 6 cm
11   two of 8 cm by 8 cm, and four of 8 cm by 2 cm   *12*   60 cm

15 two   *16* two, four, two
18a green   *b* red   *c* two, two   *d* two   *e* four

Page 124   Revision Exercise 1B

2a sphere   *b* cone   *c* cylinder   *d* pyramid   *e* cube, cuboid
3a sphere   *b* cone   *c* cylinder
4a sphere   *b* cone (on its side), cylinder (on its side)
  c cuboid, cube, pyramid, cone, cylinder
5a twelve   – *b* three   *c* cube   *d* It has two square faces.
6a yes   *b* no   *7* 52 cm   *9* 16 cm by 16 cm, or 8 cm by 8 cm
10 two cards 4 cm by 5 cm, two 4 cm by 10 cm, two 5 cm by 10 cm
13 four   *15a* (i) and (ii)   *b* (i)   *c* (i)

Page 126   Revision Exercise 2A

1a P   *b* V, N, A, Z   *c* H, I
2a 90°   *b* 36°   *c* 24°   *d* 15°
3 24   *4* 25°   *5a* 90°   *b* 150°   *c* 75°
6a 4, 1   *b* 8, 2   *c* 20, 5   *d* 18, 4½
7a between 0° and 90°   *b* 90°   *c* between 90° and 180°
  d 180°   *e* 360°
8a T   *b* T   *c* F   *d* F   *e* T   *f* F   *g* T   *h* T
9(i) 135   (ii) 15   (iii) 24
10a east   *b* south west   *c* west   *d* north
12a 90°   *b* west   *c* north west   *13* 300°
14a 115°   *b* 52°   *c* 90°   *d* (180−x)°
15a supp.   *b* neither   *c* comp.   *d* supp.   *e* comp.
16a 32°   *b* 58°   *c* 148°   *d* 328°
17a 70   *b* 17   *c* both 45   *18* 108°, 72°, 108°
19a 45°   *b* 90°   *20* 62 m   *21* 50−35 = 15 m   *22* 68 km

Page 129   Revision Exercise 2B

1a 4   *b* 8   *3a* 36   *b* 24   *c* 15   *d* 4   *e* 3
4 24   *5a* 90°, 270°   *b* 22½°, 337½°   *c* 160°, 200°
6a 8, 4   *b* 6, 3   *c* 18, 9   *7* 16   *8* 4
9a T   *b* F   *c* F   *d* T   *e* T   *f* F   *10a* 2, 4
11a 6°   *b* 12° per second
12a x must be greater than 0, less than 90
  b x must be greater than 90, less than 180
13a 000°, 180°, 225°   *b* 122°   *14a* 165°   *b* 180°
15a 135, 45   *b* 95, 85   *16* b, e   *17a* d°   *b* 135°

18 $a = c, b = d, a+d = 180, d+c = 180, c+b = 180, b+a = 180$

19a 24°   *b* 140°   *c* 90°

   d x, x, y, y make a straight angle, so x, y make a right angle

20 84 m   *21* 120°, 67–68 km

Page 132   Revision Exercise 3A

2b (2, 2), (4, 4), (8, 8)   *c* $p = q$ *3* (1, 0)

4 (2, 5), (6, 3)   *5b* Yes. No.   *c* (4, 6), (5, 8)

6b an infinite number   *c* below   *d* b is less than 4

7 the point (7, 2)   *8b* (8, 10), (9, 11), (10, 12)

Page 133   Revision Exercise 3B

2b (1, 2), (4, 8), (5, 10), (8, 16)   *c* (6, 3), (7, 2)

3a (6, 18), (7, 21)   *b* 35   *c* 108   *d* $y = 3x$

4a (7, 20), (8, 23)   *b* $p = 20$ *c* $q = 299$ *d* (2), (3), (4), (5)

5a a rectangle   *b* (11, 3), (11, 6)

6b x is greater than 10   *c* R lies to the left of the line

7b s is less than 5   *c* P lies above the line

8c the point (2, 2)

Arithmetic—Answers to Chapter 1

Page 138   Exercise 1A

1a {0, 1, 2, 3, 4, 5, 6, 7, 8, 9}   *b* {1, 3, 5, 7, 9, 11}   *c* {0, 2, 4, 6, 8, 10}

  d {1, 2, 3, 4, 5, 6, 7}   *e* {0, 1, 4, 9, 16, 25, 36, 49, 64, 81, 100}

  f {2, 3, 5, 7, 11, 13, 17, 19}   *g* {0, 1, 8, 27, 64, 125}

2a the first five whole numbers   *b* the first five odd numbers

  c even numbers from 10 to 18   *d* squares of first four whole numbers

3a even   *4a* 2, 3, 5, 7, 11, 13, 17, 19, 23, 29   *b* one   *c* one

5a 9, 11; add 2   *b* 16, 15; subtract 1   *c* 84, 86; add 2

  d 16, 25; squares of whole numbers

6a 1, 81, 256, 400, 100, 10000, 0   *b* 1, 8, 27, 0, 1000, 1000000

7a 6   *b* 28   *c* 200   *d* 31

8 whole: 0, 1, 2, 5, 8, 9, 100; natural: 1, 2, 5, 8, 9, 100; even: 0, 2, 8, 100; odd: 1, 5, 9; prime: 2, 5; square: 0, 1, 9, 100; cube: 0, 1, 8

Page 139   Exercise 1B

1a {21, 22, 23, 24}   *b* {92, 94, 96, 98}   *c* {23, 29}

  d {25, 36, 49}   *e* {0, 1, 4, 9, 16, 25, 36, 49, 64, 81, 100, 121, 144}

  f {2}   *g* {0, 1, 8, 27, 64, 125}

2a first four natural numbers   *b* squares of first three natural numbers

  c odd numbers from 101 to 109   *d* squares of 10, 11 and 12

  e prime numbers from 7 to 19

3a 0, 1, 100, 10000, 1 000 000   *b* 0, 1, 1000, 1 000 000, 1 000 000 000

4a odd

5a 10, 12; add 2   *b* 36, 49; squares of whole numbers from 2
 c 32, 64; multiply by 2   *d* 3, 4; add 1 to the third last term

6a add 4; 20   *b* squares of natural numbers; 20
 c subtract 7; 48   *d* add 2, then 3, then 4, etc; 14

7a 8   *b* 16   *8* 45; subtract 9

9a 51, 39; subtract 12   *b* 5, 7; add 1 to the second last term

Page 141   Exercise 2A

1a 0, 1, 2, 3, 4, 5   *b* 0, 1, 2, 3, 4, 5   *c* 2, 3, 4, 5, 6, 7
 d 5, 6, 7, 8, 9, 10   *e* 0, 2, 4, 6, 8, 10   *2* 15, 21, 27; 33, 39, 45

3a 2   *b* 4   *4* in the main diagonal of the square

6a even   *b* even   *c* odd

7a 14, 13   *b* 21, 12   *c* 4, 2; any values for which $a = b$

8a 11, 7   *b* 29, 21   *c* 101, 11; any values when $a = b$, also 1 and 0

9a 15   *b* 9, 5, 1; 2, 5, 8; 6, 1, 8; 2, 7, 6; 4, 3, 8

10a rows: 2, 9, 4; 7, 5, 3; 6, 1, 8   *b* 8, 1, 6; 3, 5, 7; 4, 9, 2
 c 8, 3, 4; 1, 5, 9; 6, 7, 2

Page 143   Exercise 2B

1b E, O, O   *2* 15, 21, 27; 33, 39, 45

3a 27, 27   *b* 68, 68   *c* 30, 30; yes   *4* no; when $a = b$ also 1, 0

6(i) 81   (ii) 12   *a* 68, 68   *b* 68, 68

7 all 264; M, N, O, P

Page 145   Exercise 3A

1 5   *2* 7   *3* 8   *4* 6

5 99   *6* 0   *7* 5; 7 + 5 = 12

8 12; 6 + 12 = 18   *9* 1; 4 + 1 = 5   *10* 48

11 388   *12* 395   *14a* no   *b* no, no, no

15 Take off your shoes   *16* Close the window

18 Subtract 10   *19* Spend 50 pence

Page 146   Exercise 3B

1 8   *2* 0   *3* 7   *4* 10   *5* 11   *6* 12

7 4, 9 + 4 = 13   *8* 18, 8 + 18 = 26   *9* 70, 12 + 70 = 82

10 469   *11* 949

14a false; no   *b* replacements for which $a = b$

15 Walk backward five paces

16 Turn the key anti-clockwise and open the door
18 Fly 10 km east, then 20 km south
19 Take the first turning on the right and walk up the street

Page 148   Exercise 4A

1a   0, 1, 2, 3, 4, 5   *b*   0, 3, 6, 9, 12, 15
2a   {0, 1, 4, 9, 16, 25}   *b*   the squares of the first six whole numbers
4a   0, 15, 30   *b*   45, 60, 75   *6a*   18, 12   *b*   100, 80   *c*   0, 0; no
7a   24, 24   *b*   120, 120   *c*   0, 0; * is commutative

Page 149   Exercise 4B

1   0, 15, 30; 45, 60, 75
2a   {0, 1, 2, 3, 4, 5, 6, 8, 9, 10, 12, 15, 16, 20, 25}
  b   {7, 11, 13, 14, 17, 18, 19, 21, 22, 23, 24}
3a   70, 60   *b*   35, 275   *c*   0, 110; no   *4*   true

Page 151   Exercise 5A

1   123   *2*   138   *3*   470   *4*   9800   *5*   6700
6   89700   *7*   300   *8*   2741   *9*   3150
10a   3, 11   *b*   8, 12   *c*   0, 4; no   *11a*   0   *b*   0

Page 151   Exercise 5B

1   182   *2*   109   *3*   9800   *4*   306   *5*   1956
6   2336   *7*   no. If $c = 0$ *8a*   0   *b*   0

Page 152   Exercise 6A

1   12   *2*   8   *3*   11   *4*   0   *5*   11   *6*   9
7   5, $8 \times 5 = 40$ *8*   7, $7 \times 7 = 49$ *9*   40, $2 \times 40 = 80$
10   15   *11*   37   *12*   no, no, no, no

Page 153   Exercise 6B

1   11   *2*   12   *3*   10   *4*   15, $12 \times 15 = 180$
5   6, $9 \times 6 = 54$ *6*   20, $20 \times 20 = 400$ *7*   38
8   no; values of a, b when $a = b$
9a   1, 4   *b*   1, 9   *c*   2, 50. No

Page 155   Exercise 7

1   {0, 4 8, 12, 16, 20, 24, 28}   *2*   {0, 9, 18, 27, 36, 45}
3   {0, 12, 24, 36, 48, 60, 72}   *4*   {50, 52, 54, 56, 58, 60}

5 {11, 13, 15, 17, 19}   *6* {12, 14, 16, 18}

7 { }   *8* 0, 5, 10, 15, 20, 25, 30, 35, 40, 45; last digit must be 0 or 5

9c sum of digits must be divisible by 9   *d* yes

10b sum of digits divisible by 3   *c* yes

11a {0, 2, 4, 6, 8, 10, 12, 14, 16, 18}   *b* {0, 3, 6, 9, 12, 15, 18}
  c {0, 6, 12, 18}   *d* {0, 6, 12, 18}

12a {0, 5, 10, 15, 20, 25, 30, 35}   *b* {0, 10, 20, 30}
  c {0, 15, 30}   *d* {0, 30}

Page 156   Exercise 8

1a {0, 2, 4, 6, 8, 10, 12}   *b* {0, 3, 6, 9, 12}   *c* {0, 6, 12}   *d* 6

2a {0, 3, 6, 9, 12, 15, 18, 21, 24}   *b* {0, 4, 8, 12, 16, 20, 24}
  c {0, 12, 24}   *d* 12

3a {0, 30}   *b* 30   *4a* {0, 12, 24}   *b* 12

5a {0, 40}   *b* 40

6a 6   *b* 12   *c* 30   *d* 12   *e* 40

7 15   *8* 35   *9* 10   *10* 4   *11* 6   *12* 20

13 24   *14* 24   *15* 8   *16* 20   *17* 36   *18* 36

19 18   *20* 84   *21* 14   *22* 60   *23* 12   *24* 12

25 36   *26* 30

27 {0, 4, 8, 12, 16, 20, 24, 28, 32, 36, 40, 44, 48}, {0, 6, 12, 18, 24, 30, 36, 42, 48}
   {0, 8, 16, 24, 32, 40, 48}

  a {0, 12, 24, 36, 48}   *b* {0, 24, 48}   *c* {0, 8, 16, 24, 32, 40, 48}
  d {0, 24, 48}.   LCM is 24

28 {0, 5, 10, 15, 20, 25, 30, 35, 40, 45, 50, 55, 60, 65, 70, 75},
   {0, 6, 12, 18, 24, 30, 36, 42, 48, 54, 60, 66, 72, 78}, {0, 15, 30, 45, 60, 75}

  a {0, 30, 60}   *b* {0, 30, 60}   *c* {0, 15, 30, 45, 60, 75}   *d* {0, 30, 60}
  LCM is 30

29 24 seconds; after another 24 seconds

30 60 seconds; after another 60 seconds

31 after 2 minutes; after another 2 minutes

32 1982   *33* 80000 km   *34* 30th May   *35* 3.20 pm

Page 159   Exercise 9A

2a 10   *b* {2, 3, 5, 7, 11, 13, 17, 19, 23, 29}   *3* 23, 47, 59, 101

Page 159   Exercise 9B

1a 2, 3, 5, 7, 11, 13
  b Add two odd prime numbers; add an odd and an even prime number.

2 prime: 127, 149, 151; composite: 112, 117, 145, 147   *3* 101, 103, 107, 109

4 because they will contain a number divisible by 2 (apart from 2 itself)

5a yes   *b* yes   *c* no

Page 160   Exercise 10A

1a 2, 3, 5   *b* 3, 5   *c* 2, 3, 4, 8   *d* 3, 5
2a 2, 5   *b* 3, 7   *c* 2, 13   *d* 5, 7   *e* 3, 17   *f* 7, 11
3a 3   *b* 9   *c* 12   *d* 6
4a 2, 3, 9, 11   *b* 2, 3   *c* 2, 11

Page 160   Exercise 10B

1a 2, 3, 4, 8, 9   *b* 2, 3, 4, 5, 10   *c* 2, 3, 5, 9, 10   *d* 3, 5, 11
2a 2, 47   *b* 5, 19   *c* 7, 13   *d* 2, 3, 5   *e* 3, 5, 7
3a 6   *b* 15   *c* 12   *d* 22
4a 2, 3, 5, 9, 11   *b* 2, 3, 5, 9, 11   *c* 5

Page 161   Exercise 11A

1 $2 \times 3 \times 5$ *2* $2 \times 3 \times 7$ *3* $2^2 \times 3^2$ *4* $2^3 \times 3$
5 $2^2 \times 3 \times 5$ *6* 2×3^3 *7* 2^6 *8* $3^2 \times 7$
9 $2^2 \times 5^2$ *10* $2^2 \times 3^3$ *11* $3 \times 5 \times 7$ *12* 2×3^4

Page 161   Exercise 11B

1 3×5^2 *2* $2 \times 3 \times 17$ *3* 2^7 *4* $2^3 \times 23$
5 11×19 *6* $2^2 \times 3^2 \times 5$ *7* 2×11^2 *8* 3^5
9 13×17 *10* $2^3 \times 43$ *11* $3 \times 7 \times 17$ *12* $2^6 \times 3^3$
13 $2^3 \times 3^3 \times 5$ *14* $2^2 \times 3^6$ *15* $2 \times 3^4 \times 13$ *16* $2^4 \times 3^2 \times 11$

Page 163   Exercise 12A

1 60   *2* 60   *3* 100   *4* 100   *5* 72   *6* 50
7 90   *8* 90   *9* 260   *10* 390   *11* 341   *12* 600

Page 163   Exercise 12B

1 285   *2* 413   *3* 84   *4* 2965   *5* 91   *6* 4100
7 18000   *8* 9800   *9* 596000   *10* 4100   *11* 9   *12* 1
13 1   *14* 1

Arithmetic—Answers to Chapter 2

Page 166   Exercise 1A

1 £5·73, £1·83, £0·72, £0·04, £0·06$\frac{1}{2}$ or £0·065
2 587, 225, 106, 9, 27$\frac{1}{2}$p
3 £1·91   *4* 23p   *5* 39p
6a £1·30, £13, £130   *b* £33·50, £335, £3350
 c £5·80, £58   *d* £154·80, £1548

7a £35, £3·50, £0·35   *b* £42·50, £4·25
 c £0·56, £0·28   *d* £0·08, £0·06, £0·04
8a £1·12   *b* 60p   *c* £3   *d* £5·80   *e* £590
9a £2·65   *b* 40p   *c* 27p   *d* 24p   *e* 11p
10 £2·13½

Page 167   Exercise 1B

1 £1·85   *2a* £1·17   *b* 33p   *c* 5; 3 of 10p, 2p, 1p   *3* £172
4 3½ p   *5* £66·60, £87·40   *6* £2·25
7 £29·20, £4·46   *8* £9·90   *9* 118   *10* 2p

Page 169   Exercise 2

2a cm   *b* m   *c* cm   *d* mm   *e* m
 f mm   *g* cm   *h* km
6 26 cm   *7* 24 cm
8a 16 mm, 24 cm, 44 m, 1100 m, 7 cm
 b 44 cm, 88 m, 34 m, 125 mm, 1 m 60 cm
9a 100   *b* 24   *c* 2400   *d* £288

Page 171   Exercise 3

1 40, 470, 47, 115 mm   *2* 400, 4700, 470, 473 cm
3 6, 6·8, 0·6, 25·6 cm   *4* 7, 7·5, 0·75, 2·897 m
5 8000, 19000, 19500, 1270 m
6 5, 15, 3·25, 0·7, 0·075, 0·007, 2·125, 5·005 km
7a 20·9 cm, 209 mm   *b* 29·3 cm, 293 mm
 c 12·15 m, 1215 cm   *d* 37·86 m, 3786 cm
8a 4·8 cm, 48 mm   *b* 6·3 cm, 63 mm
 c 1·41 m, 141 cm   *d* 8·92 m, 892 cm
9a 80 cm, 26 mm, 6 cm, 4 mm   *b* 2 cm, 1·48 m, 9 mm, 10·5 cm
10a 67·2 cm   *b* 7   *11* 4·9 m   *12* 32 mm
13a 1·8 m   *b* 1·5 m   *14* 11 cm, 65 cm   *15* odd, 5 cm

Page 173   Exercise 5

1b 4   *c* 4 cm²
2a 9 cm²   *b* 16 cm²   *c* 25 cm²   *d* 100 cm²   *3* no
4b yes   *c* 1 cm²   *5b* 64   *d* 64 cm²

Page 175   Exercise 6

1a 30000   *b* 25000   *2a* 5   *b* 300
3a 6   *b* 50   *4a* 4   *b* ½   *5* 16 cm²

Page 176   Exercise 7A

1a 117   *b* 1250   *c* 1350   *d* 100

2a 300 cm²   *b* 168 mm²   *c* 2250 cm²   *d* 195 km²
 e 13200 cm² = 1·32 m²   *f* 17·716 m²

3a 289 mm²   *b* 1225 m²   *c* 64 km²   *d* 625 cm²
 e 484 m²   *4* 9×4   *5* 11×6   *6* 100×53

7a 6 cm²   *b* 7 cm²   *c* 7½ cm²

8(i) 1225 cm²   (ii) 159 cm²   (iii) 468 m²   *9* 12 cm

10 9 m   *11* 12 cm   *12* 76 m

Page 178   Exercise 7B

1(i) 126 mm²   (ii) 2391 cm²   (iii) 1485 m²   (iv) 83 m²

2 12   *3a* 56   *b* 60

4a 16   *b* 45 cm   *5a* 20   *b* 4 m

6a 72   *b* 72   *7a* 45   *b* no

8 1000 cm²   *9* 125 cm   *10* 24 m², £4·80   *11* 150 cm²

12 800 cm²   *13a* 36 cm   *b* 80 m

Page 182   Exercise 8A

1a 160 mm³   *b* 648 cm³   *c* 20 litres   *d* 432 m³
 e 294 cm³   *f* 15 litres   *2* 120 cm³   *3* 60 m³

 4 17½   *5* 125 cm³, 125 ml

6a 2400 cm³   *b* 2400 ml   *c* 2·4 litres   *7* 12

8a 50 litres   *b* 5   *9* 28   *10* 50

11a 27 cm³   *b* 3 cm   *c* 37   *12* 2 cm

13a 160 mm³   *b* 8000 mm³, 8 cm³

14a 12500 cm³   *b* 1000000 cm³

Page 183   Exercise 8B

1a 576   *b* 400   *c* 1500   *d* 5   *e* 2·5

2a 420   *b* 1400   *c* 0·1

3a 16 min   *b* 24 min; no, yes   *4* 21·6 litres, 6 cm

5 1200 cm³, 2 cm; 80 cm and 3 cm

6a 64 unit³, 4 units   *b* 1000, 10   *c* 9261, 21   *7* 7 cm, 2

8a 110, 112, 128, 132, 140, 147   *b* 518 cm³   *9* 200000 kg

Page 185   Exercise 9

1 4, 2·872, 0·479, 0·075 kg   *2* 2500, 105, 3, 250, 500, 750 g

3 5, 9·81, 0·5, 0·025 g   *4* 7000, 7500, 7520, 75 mg

5 1·9 kg   *6* 3·125 kg, £2·75   *7* 18·7 kg, £21·25

8   1·36 kg   9   10·75 kg   10a   130   b   £3·12, 12½p
11   102·5 kg   12   5284 kg

Arithmetic—Answers to Chapter 3

Page 189   Exercise 1

2   3 cm, 4 cm, 6 cm, 2 cm, 1½ cm, 8 cm, 10 cm, 1⅕ cm or 1·2 cm

3    $\frac{5}{7}$ 4 $\frac{1}{12}, \frac{3}{12}(\frac{1}{4}), \frac{2}{12}(\frac{1}{6})$ 5 $\frac{4}{12}(\frac{1}{3})$ 6 $\frac{5}{26}, \frac{21}{26}$

9    $\frac{1}{15}, \frac{7}{15}, \frac{10}{15}(\frac{2}{3})$ 10 $\frac{1}{4}, \frac{1}{2}, \frac{3}{4}$ 11 $\frac{1}{60}, \frac{30}{60}(\frac{1}{2}), \frac{1}{3600}, 1$

12a   2, 5, 1, 9   b   5, 4, 8, 100

Page 190   Exercise 2A

3    $\frac{2}{10}, \frac{3}{15}$, etc.   4 $\frac{4}{6}, \frac{6}{9}$, etc.   5 $\frac{10}{18}, \frac{15}{27}$, etc.   6 $\frac{2}{3}, \frac{6}{9}$, etc.

7    $\frac{2}{3} = \frac{6}{9} = \frac{8}{12} = \frac{12}{18} = \frac{20}{30} = \frac{100}{150}$ 8a $\frac{1}{3}$ b $\frac{2}{3}$

9a $\frac{2}{3}$ b $\frac{3}{4}$ 10a $\frac{1}{2}$ b $\frac{4}{5}$ 11a $\frac{3}{5}$ b $\frac{4}{5}$

12a $\frac{3}{5}$ b   1   13a $\frac{2}{3}$ b $\frac{6}{7}$ 14a $\frac{1}{2}$ b   1

15a $\frac{1}{3}$ b $\frac{2}{3}$ 16 $\frac{1}{3}$ 17 $\frac{4}{5}$ 18 $\frac{1}{2}$ 19 $\frac{7}{8}$

21    $\frac{1}{4}, \frac{1}{3}, \frac{3}{4}, \frac{2}{5}$ 22 $1\frac{4}{5}$ 23 $3\frac{2}{3}$ 24 $3\frac{3}{4}$ 25 $5\frac{3}{4}$

26    $5\frac{2}{5}$ 27 $6\frac{1}{9}$ 28 $6\frac{9}{11}$ 29 $1\frac{3}{4}$ 30 $\frac{11}{4}$ 31 $\frac{11}{2}$

32    $\frac{23}{3}$ 33 $\frac{18}{5}$ 34 $\frac{101}{10}$ 35 $\frac{9}{8}$ 36 $\frac{24}{5}$ 37 $\frac{137}{12}$

38    $\frac{2}{4}, \frac{3}{6}$, etc.   39 $\frac{3}{5}, \frac{6}{10}$, etc.   40 $\frac{6}{8}, \frac{9}{12}$, etc.

41    $\frac{38}{40}, \frac{57}{60}$, etc.

Page 192   Exercise 2B

3a $\frac{7}{12} = \frac{14}{24} = \frac{35}{60} = \frac{21}{36} = \frac{70}{120} = \frac{7x}{12x}$ b   7, 14, 35, 21, 70, 7x

4    $\frac{5}{6}$ 5 $\frac{2}{3}$ 6 $\frac{2}{5}$ 7 $\frac{2}{5}$ 8 $\frac{2}{3}$ 9 $\frac{3}{4}$

10    $\frac{3}{7}$ 11 $\frac{17}{25}$ 12 $\frac{5}{7}$ 13 $\frac{2}{3}$ 14 $\frac{a}{b}$ 15 $1\frac{3}{5}$

16    $3\frac{1}{5}$ 17 $3\frac{5}{8}$ 18 $2\frac{7}{12}$ 19 $6\frac{5}{9}$ 20 $33\frac{1}{3}$ 21 $\frac{5}{4}$

22    $\frac{16}{3}$ 23 $\frac{77}{8}$ 24 $\frac{135}{8}$ 25 $\frac{125}{3}$ 26 $\frac{4}{10}, \frac{6}{15}$, etc.

27    $\frac{5}{16}, \frac{15}{48}$, etc.   28 $\frac{8}{14}, \frac{12}{21}$, etc.   29 $\frac{7}{20}, \frac{14}{40}$, etc.

30    $\frac{2x}{2y}, \frac{3x}{3y}$, etc.   31 $\frac{1}{2}, \frac{2}{4}$, etc.   32 $\frac{6}{10}, \frac{3}{5}$, etc.

33    $\frac{25}{100}, \frac{1}{4}$, etc.   34 $\frac{2x}{4}, \frac{3x}{6}$, etc.   35 $\frac{4}{2x}, \frac{6}{3x}$, etc.

36    $\frac{45}{65}, \frac{54}{78}$

Page 194   Exercise 3A

1    $\frac{2}{2}, \frac{3}{3}$, etc.   2 $\frac{20}{2}, \frac{30}{3}$, etc.   3 $\frac{50}{2}, \frac{75}{3}$, etc.   4 $\frac{2}{4}, \frac{5}{10}$, etc.

5    $\frac{3}{4}$, etc.   6 $\frac{1}{2}$, etc.   7 $\frac{4}{6}(\frac{2}{3})$, etc.   8 $\frac{3}{8}$, etc.

9 $\frac{3}{10}$, $\frac{4}{10}$, etc.   *10* $1\frac{1}{16}$, etc.   *11* $\frac{1}{20}$, etc.   *12* $\frac{7}{24}$, etc.

13 $\frac{1}{4}$, $\frac{1}{2}$, $\frac{3}{4}$ *14* $\frac{3}{8}$, $\frac{1}{2}$, $\frac{5}{8}$ *15* $\frac{1}{2}$, $\frac{2}{3}$, $\frac{3}{4}$ *16* $\frac{2}{5}$, $\frac{1}{2}$, $\frac{3}{5}$

17a $\frac{4}{9}$, $\frac{7}{18}$, etc.   *b* $\frac{1}{6}$, $\frac{2}{15}$, etc.   *18* $\frac{5}{8}$, $\frac{3}{4}$, $\frac{7}{8}$, etc.

Page 194   Exercise 3B

1 $\frac{6}{2}$, $\frac{9}{3}$, $\frac{12}{4}$, etc.   *2* $\frac{30}{2}$, $\frac{45}{3}$, $\frac{60}{4}$, etc.   *3* $\frac{200}{2}$, $\frac{300}{3}$, $\frac{400}{4}$, etc.

4 $\frac{6}{8}$, $\frac{9}{12}$, $\frac{12}{16}$, etc.   *5* $\frac{2x}{2}$, $\frac{3x}{3}$, $\frac{4x}{4}$, etc.   *6* $\frac{2}{3}$, etc.

7 $\frac{1}{2}$, etc.   *8* $\frac{9}{40}$, etc.   *9* $\frac{1}{16}$, etc.   *10* $\frac{1}{20}$, etc.

11 0·5, etc.   *12* 0·45, etc.   *13* $\frac{5}{6}$, etc.   *14* $\frac{1}{2}$, $\frac{1}{4}$, $\frac{1}{8}$

15 1, $\frac{11}{12}$, $\frac{7}{8}$ *16* $1\frac{5}{6}$, $1\frac{3}{4}$, $1\frac{2}{3}$ *17* $\frac{4}{5}$, $\frac{7}{10}$, $\frac{69}{100}$ *18* $\frac{5}{6}$, $\frac{19}{24}$

19 $\frac{59}{300}$, $\frac{29}{150}$, $\frac{19}{100}$, $\frac{14}{75}$, $\frac{11}{60}$, $\frac{9}{50}$, $\frac{53}{300}$, $\frac{13}{75}$, $\frac{17}{100}$, etc.

Page 195   Exercise 4A

1 £1·58   *2* £0·70   *3* 35p   *4* 750 ml   *5* 50 min

6 70 cm   *7* 250 g   *8* 750 g   *9* £4·97

10 $\frac{1}{90}$, $\frac{2}{9}$, $\frac{17}{18}$, $\frac{1}{180}$ *11a* $\frac{4}{5}$ *b* $\frac{1}{5}$ *c* $\frac{16}{25}$

12a £1·76   *b* £1·80   *c* £1·92   *d* £2   *13* 60, $22\frac{1}{2}$, 135

14 £21·50   *15a* 108 g   *b* 130 g   *17* 1 h 20 min

18 368, 184, 230   *19* 820 km   *20a* $\frac{3}{10}$ *b* 110 m

Page 195   Exercise 4B

1 42p   *2* £3·08   *3* 25 min   *4* 70 cm   *5* 45 min

6 56 kg   *7a* 45°   *b* $67\frac{1}{2}$°   *c* 150°   *d* 315°

8 $\frac{1}{9}$, $\frac{7}{18}$, $\frac{4}{45}$, $\frac{1}{8}$, $\frac{1}{120}$ *9a* $\frac{3}{4}$ *b* $1\frac{3}{4}$ *c* $3\frac{3}{4}$

10 $\frac{1}{3}$, $1\frac{1}{3}$, $3\frac{1}{3}$ and $\frac{5}{8}$, $1\frac{5}{8}$, $3\frac{5}{8}$ *11a* $\frac{1}{5}$ *b* $\frac{1}{6}$

12a £263·20   *b* £789·60   *13* yes   *14* £16·80, $\frac{7}{10}$

15a $\frac{2}{5}$ *b* $\frac{1}{2}$

Page 198   Exercise 5A

1 6   *2* 24   *3* 35   *4* 24   *5* 72   *6* 30

7 12   *8* 40   *9* $\frac{5}{6}$ *10* $1\frac{1}{15}$ *11* $\frac{17}{24}$ *12* $1\frac{5}{12}$

13 $\frac{3}{4}$ *14* $1\frac{1}{12}$ *15* $\frac{1}{2}$ *16* $3\frac{8}{15}$ *17* $5\frac{17}{20}$ *18* $1\frac{7}{20}$

19 $5\frac{5}{14}$ *20* $6\frac{1}{12}$ *21* $8\frac{3}{20}$ *22* $2\frac{1}{8}$ *23* $11\frac{7}{10}$ *24* $1\frac{3}{10}$

26 $\frac{4}{15}$ *27* 240

28a

1	$3\frac{1}{2}$	3
$4\frac{1}{2}$	$2\frac{1}{2}$	$\frac{1}{2}$
2	$1\frac{1}{2}$	4

b

$\frac{5}{8}$	$1\frac{1}{4}$	$\frac{3}{8}$
$\frac{1}{2}$	$\frac{3}{4}$	1
$1\frac{1}{8}$	$\frac{1}{4}$	$\frac{7}{8}$

29a $1\frac{3}{4}$ *b* $1\frac{5}{6}$ *c* $3\frac{7}{10}$ *d* $\frac{9}{10}$

Page 199   Exercise 5B

1	$1\frac{7}{24}$	**2**	$1\frac{1}{8}$	**3**	$4\frac{1}{12}$	**4**	$1\frac{1}{12}$	**5**	$3\frac{5}{12}$	**6**	$2\frac{7}{8}$
7	3	**8**	$7\frac{3}{10}$	**9**	$\frac{13}{24}; \frac{1}{96}$	**10**	$3\frac{7}{8}$	**11**	$\frac{13}{16}$	**12**	$\frac{11}{12}$
13	$\frac{35}{48}$	**14**	14	**15**	$3\frac{1}{4}$	**16**	$7\frac{1}{5}$	**17**	6		

18a $7\frac{11}{12}$ **b** $4\frac{11}{12}$ **c** 26   **19** $\frac{1}{6}$

20a $\frac{7}{48}$ **b** £5760, £2400, £2520   **21a** A   **b** no   **c** $\frac{13}{60}$

22 each $= 1\frac{3}{8}$ **23** $7\frac{7}{8}, 9\frac{1}{4}, 10\frac{5}{8}$ **24** each $= 1\frac{3}{8}$ **25** each $= 2\frac{3}{4}$

Page 202   Exercise 6A

1a $\frac{2}{3}$, 0, 2, $2\frac{2}{3}$, 4, 12   **b** 5, $1\frac{1}{3}$, 20, 0, $2\frac{1}{2}$, $4\frac{1}{6}$ **c** 10, $17\frac{1}{2}$, 0, 35, $7\frac{1}{2}$

2	$\frac{1}{6}$	**3**	$\frac{3}{8}$	**4**	5	**5**	$3\frac{1}{3}$	**6**	$\frac{3}{4}$	**7**	6
8	$4\frac{1}{2}$	**9**	0	**10**	2	**11**	2	**12**	3		
13	3	**14**	30 cm	**15**	16 cm	**16**	$55\frac{1}{4}$ cm², $12\frac{15}{16}$ cm²				

17 18 cm, $20\frac{1}{4}$ cm²   **18** 5 m, $1\frac{9}{16}$ m²

Page 202   Exercise 6B

1a $\frac{3}{8}$, 0, 3, $\frac{3}{4}$, $1\frac{1}{8}$, 6   **b** 5, $2\frac{1}{2}$, 0, $7\frac{1}{2}$, 10, 45   **c** 28, $9\frac{1}{3}$, 56, 0, $2\frac{1}{3}$, 7

2	$\frac{1}{2}$	**3**	$\frac{1}{2}$	**4**	$\frac{5}{6}$	**5**	$3\frac{1}{2}$	**6**	26	**7**	$\frac{1}{4}$
8	$\frac{1}{8}$	**9**	$\frac{1}{24}$								

10a $\frac{7}{24}$ **b** $\frac{5}{24}$ **c** $\frac{1}{12}$ **d** $\frac{3}{8}$ **11a** $\frac{1}{6}$ **b** $\frac{2}{5}, \frac{4}{15}, \frac{2}{15}$

12a $\frac{4}{15}$ **b** $\frac{2x}{15}$ **c** $\frac{3}{4x}$ **d** $\frac{pq}{6}$ **e** $\frac{3}{4}$ **f** $1\frac{1}{2}$ **g** 1   **h** 1

13	$1\frac{5}{7}$	**14**	$3\frac{3}{10}$	**15**	$\frac{5}{9}$	**16**	7	**17**	$11\frac{1}{4}$	**18**	$27\frac{1}{2}$
19	$13\frac{1}{2}$	**20**	$\frac{11}{12}$	**21**	$5\frac{7}{8}$						

Page 204   Exercise 7A

1	$1\frac{1}{2}$	**2**	$2\frac{1}{2}$	**3**	$2\frac{1}{3}$	**4**	$\frac{4}{9}$	**5**	$1\frac{1}{3}$	**6**	2
7	$1\frac{1}{6}$	**8**	$\frac{1}{8}$	**9**	$1\frac{1}{6}$	**10**	$\frac{5}{6}$	**11**	4	**12**	$\frac{2}{3}$
13	$2\frac{1}{4}$	**14**	$1\frac{2}{3}$	**15**	$1\frac{5}{9}$	**16**	4	**17**	$1\frac{1}{3}$	**18**	$5\frac{1}{3}$
19	$\frac{2}{5}$	**20**	10	**21**	$\frac{1}{6}$			**22**	$\frac{1}{4}$		
23	$\frac{5}{12}$			**24**	$\frac{2}{3}$			**25**	+ or ÷		
26	−	**27**	×	**28**	+	**29**	÷	**30**	−	**31**	×
32	× or −			**33**	÷	**34**	+	**35**	×	**36**	÷
37	−	**38**	+								

Page 205   Exercise 7B

1	$1\frac{1}{4}$	**2**	2	**3**	$2\frac{5}{8}$	**4**	$4\frac{4}{5}$	**5**	$1\frac{1}{3}$	**6**	$2\frac{5}{8}$
7	$1\frac{2}{7}$	**8**	$\frac{8}{11}$	**9**	$1\frac{1}{2}$	**10**	$2\frac{7}{16}$	**11**	$\frac{5}{8}$		
12	$\frac{1}{5}$			**13**	$\frac{1}{5}$			**14**	$\frac{5}{6}$		**15**	$\frac{5}{8}$
16	−	**17**	×	**18**	+	**19**	÷	**20**		**21**	×

22 ×   *23* ÷   *24* −

25 $\frac{1}{2}+\frac{1}{4} = \frac{3}{4}, \frac{3}{4}-\frac{1}{2} = \frac{1}{4}, \frac{3}{4}-\frac{1}{4} = \frac{1}{2}$

26 $\frac{2}{3}\times\frac{3}{4} = \frac{1}{2}, \frac{1}{2}\div\frac{2}{3} = \frac{3}{4}, \frac{1}{2}\div\frac{3}{4} = \frac{2}{3}$

27 12   *28* $5\frac{1}{3}$ *29* $\frac{3}{16}$ *30* 2   *31* 16   *32* 4

33 $\frac{1}{9}$ *34* $1\frac{1}{3}$ *35* 5

Page 206   Exercise 8A

1 7:8   *2* 1:5   *3* 1:2   *4* 2:3   *5* 1:8   *6* 1:20

7 4:5   *8* 4:3   *9* 1:3   *10* 12, 30, 24, 40, $1\frac{1}{2}$, $3\frac{3}{4}$

11a 2:1   *b* 2:1   *c* 4:1   *12a* 4:3   *b* 16:9

13 4:1   *14* 1:4, 1:3, 1:2, 2:3   *15* 6:5   *16* 5:7   *17* 5:8

Page 207   Exercise 8B

1 7:18   *2* 5:12   *3* 6:5   *4* 19:40   *5* 2:3   *6* 4:9

7 23:40, 5:6, 35:52, 3:5, 4:3   *8* 3:2, speeds 2:3   *9* 13:15

10 4:3   *11* 5:3, 5:2, 2:3   *12* $2\frac{1}{2}$ km

13 1:200   *a* 3 cm   *b* 8 cm

14 1250 m, 750 m   *15* 4·5 m   *16a* 1:2   *b* 1:4   *c* 1:8

17a 2:3   *b* 4:9   *c* 8:27

Page 209   Exercise 9

1 $\frac{1}{5}$ *2* $\frac{3}{10}$ *3* $\frac{2}{5}$ *4* $\frac{3}{20}$ *5* 1   *6* $\frac{1}{2}$

7 $1\frac{3}{4}$ *8* $\frac{12}{25}$ *9* $\frac{4}{25}$ *10* $\frac{6}{25}$ *11* $\frac{1}{50}$ *12* $\frac{1}{8}$

13 $\frac{9}{20}$ *14* $1\frac{1}{20}$ *15* $\frac{19}{20}$ *16* $\frac{1}{40}$ *17* $\frac{1}{3}$ *18* $1\frac{1}{2}$

19 $\frac{2}{3}$ *20* $\frac{7}{8}$ *21* £6   *22* $52\frac{1}{2}$, £175   *23* 44

24 £3, $1\frac{1}{4}$ kg   *25* 12 m   *26* 200 m, £4   *27* $\frac{1}{2}$

28 $19\frac{1}{5}$ m, £3·20   *29* $\frac{1}{2}$ litre   *30* 2 hours, £5·50   *31* £2·34

32 903   *33* £23·40   *34* £5   *35* $22\frac{1}{2}$ cm

36 £2·10   *37* 125 ml   *38* 8 km   *39* £13·57   *40* 70 m²

41 30 kg   *42* 160 g   *43* 3p   *44* $6\frac{1}{2}$p   *45* £1·23

Page 210   Exercise 10

1 50%   *2* 25%   *3* 40%   *4* 75%   *5* 80%   *6* 6%

7 76%   *8* $12\frac{1}{2}$%   *9* 100%   *10* 250%   *11* 1%   *12* $8\frac{1}{3}$%

13a $42\frac{6}{7}$%   *b* $57\frac{1}{7}$%   *14* $93\frac{1}{3}$%   *15* $33\frac{1}{3}$%

16 $83\frac{1}{3}$%   *17* 80%   *18* 25%   *19* 70%

20 $33\frac{1}{3}$%   *a* $12\frac{1}{2}$%   *b* $16\frac{2}{3}$%   *c* $13\frac{8}{9}$%

Page 211   Exercise 11

1 £25·20   **2** £324, £396   **3** £2·34

4 saves £369; spends £1431, £225, £450, £144, £180, £432

5 £780000, £819000   **6** 1260 km/h   **7a** $12\frac{1}{2}\%$ **b** $14\frac{2}{7}\%$

8 75 cm², 144%, 44%   **9** $82\frac{1}{2}\%$

Page 214   Topic to explore

1a even   **b** even   **c** odd

2a O   **b** O   **c** yes   **d** commutative law
e O + E = O   **f** E + O = O   **g** yes   **h** associative law

3a E   **b** E   **c** yes   **d** commutative law
e E × E = E   **f** E × E = E   **g** yes   **h** associative law
i E + E = E   **j** E × O = E   **k** yes   **l** distributive law

4a The operation is 'Double the first number and multiply by the second'.
c yes

Arithmetic—Answers to Revision Exercises

Page 216   Revision Exercise 1A

1a 0, 1, 64, 121, 225, 900, 2500, 10000   **b** 0, 1, 27, 1000

2a {0, 1, 2, 3, 4, 5, 6, 7, 8, 9, 10, 11}   **b** {50, 52, 54, 56, 58, 60, 62}
c {0, 1, 4, 9, 16, 25, 36, 49, 64, 81}   **d** {2}

3a 29, 36   **b** 17, 23   **c** 34, 66   **d** 96, 95   **e** 41, 51   **f** 4, 2

4a 6, 0   **b** 16, 22   **c** 16   **d** 27, 125

5

11	6	13
12	10	8
7	14	9

a

4	9	6	15
5	16	3	10
11	2	13	8
14	7	12	1

b

8	3	4
1	5	9
6	7	2

c

6a 11, 10   **b** 5, 10   **c** 11, 17   **d** 13, 23; no, no

7a 14   **b** 6   **c** 30   **d** 30   **e** 60   **8** 31, 37

9a 2×3^2 **b** $2 \times 3 \times 5$ **c** $2^2 \times 3 \times 7$ **d** 3×5^2

10a 149   **b** 122   **c** 990   **d** 5300   **e** 100   **f** 320

Page 217   Revision Exercise 1B

1a 100, 625, 10000, 1000000
b 0, 1, 125, 1000, 1000000, 1000000000

2a {101, 102, 103, 104, 105, 106, 107, 108, 109}   *b* {91, 93, 95, 97, 99}
  c {0, 1, 4, 9, 16, 25, 36, 49, 64, 81, 100, 121, 144, 169, 196}
  d {41, 43, 47}

3a 23, 28   *b* 7, 0   *c* 64, 125   *d* 28, 39   *e* 5, 5   *f* 30, 42
4a 101, 11   *b* 9, 3
5a 25   *b* 21   *c* 42   *d* 36   *e* 54
  f 48   *g* 8. No
6a 18   *b* 12   *c* 0   *d* 0   *e* 16
  f 4   *g* 100   *h* 2500. No; no
7a 60   *b* 6   *c* 12   *d* 60   *e* 60
8 105   *9a* $3^2 \times 7$ *b* $2^4 \times 3$ *c* $2^2 \times 3^2 \times 5 \times 7$
10a 129   *b* 1300   *c* 17000   *d* 120   *e* 1380   *f* 3700
11 111, 222, 333, 444, ..., 999
12 1, 121, 12 321, ..., 12 345 654 321
13a 381   *b* 169   *c* 254   *14b* 27450   *c* 39 245, 293 680
15a 122 400   *b* 196 550   *c* 94 775
16 *a, b, d, e*   *17* 28   *18* 3, 11
19 $(193 \times 193) - (192 \times 194) = 1$; 998 001
20a T   *b* F   *c* T   *d* F   *e* F   *f* T

Page 220   Revision Exercise 2A

1a 60p   *b* £2·12½   *c* 90p   *d* £2·01   *e* £4·48   *f* £10·11½
2 14p   *3a* 40 mm   *b* 46 cm   *c* 1850 m   *4* 105·6 m
5 975 m   *6* 4   *7* 25
8a 1250 mm   *b* 125 cm   *c* 1·25 m
9a 52 mm   *b* 18 m   *10a* 1·4 m   *b* 6 m
11 48   *12* £2100   *13* About 78
14a 320 cm²   *b* 324 mm²   *c* 340 m²   *d* 510 mm²   *e* 117 km²
15 £11 520   *16* £735   *17* 94 cm²; 163 m²
18a 1000, 1 000 000   *b* $\frac{1}{250}$ *c* $\frac{1}{4}$ *19a* 90 cm³   *b* 4 m³
20a 1000 cm³   *b* 512 m³   *c* 125 mm³   *d* 15 625 m³
21a 7·5 litres   *b* 9 litres   *22* 4 m
23a 1·8 kg, £1·56   *b* 3·6 kg, £1·92

Page 222   Revision Exercise 2B

1 £117·57, 1½p   *2* about 8 km   *3* 202 cm   *4* 73 km² approx.
5a 1540 m²   *b* 77   *c* £69·30   *6* £71·80   *7* 3 m
8 120   *9a* 48   *b* 36   *c* 15   *10* $\frac{2}{5}$ *11* 2000
12a 47 m²   *b* 30 m³   *13* 21 000 kg
14 140, 128, 147, 112, 132, 110   *15* 530 kg   *16* 8000 tonnes

Answers

Page 224   Revision Exercise 3A

1a $\frac{1}{2}$ b $\frac{5}{6}$ c $\frac{4}{5}$ d $\frac{4}{9}$ e 1

2a £316   b 24 minutes   c 350 ml

3a £580   b 1 h 45 min   c £5·76   4 £1·08$\frac{1}{2}$

5a 12   b 6   c 10   d 12   6 £7·87$\frac{1}{2}$

7a $\frac{11}{12}$ b $\frac{1}{24}$ c $5\frac{1}{4}$ d $2\frac{2}{15}$ 8a $\frac{1}{6}$ 9$\frac{1}{3}$ c $1\frac{1}{6}$ $2\frac{1}{4}$

9a 30 m³   b $40\frac{1}{2}$ mm³   c $13\frac{1}{2}$ cm³   d $26\frac{1}{4}$ m³

10a 42 cm   b $73\frac{1}{2}$ cm²   11a $3\frac{1}{2}$, 4   b $\frac{5}{8}$, $1\frac{1}{16}$

12a $3\frac{9}{16}$ b $\frac{3}{16}$ 13a 3 : 7   b 9 : 20   14 4 : 3

15a 2 : 3   b 4 : 9   c 8 : 27

16a $\frac{3}{10}$, 30%   b $\frac{3}{200}$, $1\frac{1}{2}$%   c $\frac{19}{20}$, 95%   d $\frac{3}{4}$, 75%   e $\frac{7}{8}$, $87\frac{1}{2}$%

17a 25p   b £60   c 9p   18 1150   19 £13·95

20 £25·80   21 $43\frac{3}{4}$%, $56\frac{1}{4}$%

22a 10%, $11\frac{1}{9}$%   b 90%, $88\frac{8}{9}$%   23 $15\frac{4}{15}$%

24 $53\frac{1}{3}$%   25a each = $1\frac{5}{12}$ b $\frac{1}{4}$ and $\frac{1}{8}$

26 0, $2\frac{2}{3}$, $5\frac{1}{3}$, 8, $10\frac{2}{3}$, $13\frac{1}{3}$, 16, $18\frac{2}{3}$, $21\frac{1}{3}$, 24, $26\frac{2}{3}$, $29\frac{1}{3}$, 32

Page 226   Revision Exercise 3B

1a $1\frac{9}{20}$ b $\frac{19}{48}$ 2a $112\frac{1}{2}$°   b $3\frac{1}{2}$ c £3·99

3a 67·5 cm   b 2970 ml, or 2 litres 970 ml   c £36·63

4 £1·07   5 £4·07   6 60000   7 £648   8 11   9 $24\frac{1}{2}$ kg

10 66p   11 50 km/h   12a 14   b $\frac{7}{36}$ c 0   d $\frac{7}{8}$

13 1240 cm³   14a 1 : 3   b 2 : 3   c 1 : 100000

15a 2 : 3   b 11 : 26   c 1 : 4

16a $\frac{1}{3}$, $33\frac{1}{3}$%   b $\frac{1}{5}$, 20%   c $\frac{1}{8}$, $12\frac{1}{2}$%   d $\frac{1}{80}$, $1\frac{1}{4}$%

17 $32\frac{1}{2}$ kg, 156 kg   18 18p

19 $37\frac{1}{2}$%   a 15   b 3   c 9   20 72%

21 690 kg   22 between 48 mm and 52 mm   23 $11\frac{1}{9}$%, $88\frac{8}{9}$%

24a 1 : 2   b 1 : 4   c 1 : 64   d 16 : 1   e 4 : 1   f 1 : 32

26a $\frac{5}{6}$ b $2\frac{2}{5}$ c $1\frac{5}{12}$ 27a $1\frac{1}{4}$ b $\frac{3}{4}$ c $1\frac{5}{8}$ d $\frac{5}{6}$

Page 228   'True-False' Revision Exercise

1 T   2 F   3 F   4 T   5 T   6 T   7 F

8 F   9 T   10 F   11 F   12 T   13 T   14 F

15 F   16 T   17 F   18 F   19 F   20 T   21 T

22 F   23 T   24 F   25 F